Praise for Dr. Jian Wang and Being Online

"The first time I met Dr. Jian Wang (we call him the Doctor), I was shocked by his vision of the future of internet technology. I wish we had met earlier."
—Jack Ma, cofounder and former executive chairman of Alibaba Group, from his foreword

"In *Being Online* Jian Wang offers an expert view into the future of information and interconnectivity. The convergence of digital technologies promises to transform the economy and the society in ways that will affect every company and every individual. This book provides a roadmap for managers, technologists, entrepreneurs, and ordinary citizens."
—Mauro F. Guillén, dean, Cambridge Judge Business School

"Jian Wang was an early star of Microsoft's pioneering research lab in Beijing, Microsoft Research Asia. From there, he went on to greater heights, becoming a driving force in Alibaba's success story. Few people in the world have his perspective of our web-enabled future—spanning eastern and western cultures. *Being Online* offers unique and valuable insights into the cloud and our increasingly interconnected world. It is not to be missed."
—Robert Buderi, journalist, entrepreneur, author of *Engines of Tomorrow, The Invention That Changed the World,* and other books and former editor in chief of *MIT Technology Review*

"*Being Online* reads like an encyclopedia of technology knowledge at your disposal. Dr. Wang's book can serve as a launching pad for the creative technical minds that will build the essential tools for our future."
—Ed Bennett, former president and CEO of VH1 and former CEO of Prodigy

"Dr. Jian Wang was, and is, a moonshot thinker. From Alibaba Cloud to City Brain, his own brain seemingly knows no limitation, and he puts it to the task in this book by distilling complex ideas into approachable frameworks, philosophies, and tangible examples. True to form, Dr. Wang doesn't hold back on making bold observations or predictions. This book is for those who are curious about how the online world came to be and interested in where it is going."
—K. Guru Gowrappan, CEO of Verizon Media/Yahoo!

"Jian Wang presents an extremely knowledgeable and compelling account of our emerging online lives. His exposition of the City Brain initiative is particularly inspiring for the future of smart cities. I thoroughly recommend reading this book."
—Peter G. Rowe, Raymond Garbe Professor of Architecture and Urban Design and Harvard University Distinguished Service Professor

"Whatever your background, this fascinating book will likely enrich your understanding of the potential for data once it is interactively available online. Wang believes being online will be the 'essence of this era' and that our understanding of it has just begun. At the highest level, having real-time interactive

access to data, knowledge, partners, and users is ultimately a major amplifier of innovation. He persuasively makes the case that the 'new continent' he describes will be constantly expanding in valuable new ways."
—Curt Carlson, former chairman and CEO of SRI

"Through his cross-disciplinary training and working experiences, Dr. Wang gained a unique and profound understanding of data, cloud computing, and the internet. 'Being online' is what makes 'big data' interesting and valuable. This book will help you think deeper and further about the future of digitization and its impact on our societies. It's a must-read for anybody who cares about data and its potential business opportunities."
—Zuo-Jun (Max) Shen, Chancellor's Professor, Industrial Engineering and Operations Research & Civil and Environmental Engineering, UC Berkeley; vice president and pro-vice-chancellor (research), University of Hong Kong

"When I first met Dr. Jian Wang ten years ago, he couldn't stop talking about the exciting future of cloud computing. Today, he has become a pioneer and doer of Alibaba's fundamental technology infrastructure. In the book, he not only shares his unique experience, but also makes bold predictions for the future. I believe that no matter what industry you are in, this wonderful book will open your mind."
—Joseph Tsa, executive vice president, Alibaba Group

"*Being Online* can greatly inspire the younger generation. In it, the Doctor generously shares the stories of YunOS and

the 'De-IOE' of Aliyun (Alibaba Cloud), which demonstrate both the reach of his ambition and his persistence. Reading his book is knowing the person."
—Ming-kai Tsai, chairman of the board and former CEO, MediaTek Inc.

"You don't really understand the internet if you don't understand what being online is. At the World Internet Conference in Wuzhen, I was so excited after listening to Dr. Jian Wang speak that I sent a text message to Jack Ma saying, 'Jian Wang is Alibaba's treasure!' Jack replied, 'He is China's treasure.'"
—Shuo Qin, former editor in chief of *China Business News Daily*

BEING
ONLINE

BEING ONLINE

On Computing,
Data, the Internet,
and the Cloud

JIAN WANG

FOREWORD BY JACK MA

Arcade Publishing • New York

First English-language Edition

Arcade Publishing books may be purchased in bulk at special discounts for sales promotion, corporate gifts, fund-raising, or educational purposes. Special editions can also be created to specifications. For details, contact the Special Sales Department, Arcade Publishing, 307 West 36th Street, 11th Floor, New York, NY 10018 or arcade@skyhorsepublishing.com.

Arcade Publishing® is a registered trademark of Skyhorse Publishing, Inc.®, a Delaware corporation.

Visit our website at www.arcadepub.com.

10 9 8 7 6 5 4 3 2 1

Library of Congress Cataloging-in-Publication Data is available on file.
Library of Congress Control Number: 2021939597

Editorial consultant: Weihua Liu
Cover design by Erin Seaward-Hiatt

ISBN: 978-1-951627-79-9
Ebook ISBN: 978-1-951627-96-6

Printed in the United States of America

To my parents for a lifetime of loving and caring

To my wife and daughter for your love and lovely encouragement

Contents

Foreword

by Jack Ma

The first time I met Dr. Jian Wang (we call him the Doctor), I was shocked by his vision of the future of internet technology. I wish we had met earlier.

The first time I heard the Doctor speak, during a strategy meeting at the company, about the future era of data, I was impressed by his understanding of data technology and his determination, which has led to today's strategic planning at Alibaba.

The first time I heard about the Doctor's proposal for the research and development of YunOS [the mobile operating system], I was surprised and even pissed off at how bold he was.

The first time I heard people complain about how awful the YunOS cell phone was, I witnessed the tenacity of the doctor and his team.

When, for the first time, the dispute occurred between Hadoop [a software platform for developing and running large-scale data processing] and Cloud Ladder technology [code name for data processing software that was part of Aliyun's Apsara cloud operating system], I was pleasantly surprised to see the scientific spirit and persistence of an amazing and absolutely responsible technical leader.

Everyone has criticized and evaluated the ways the Doctor manages, communicates, and executes, but no matter what, what we see is always a humble, pure, and childlike grown-up. The Doctor is a human and not a god. Everyone knows his shortcomings, which I also knew quite a lot about. But his great merits are probably visible to only a very few.

If we had had the Doctor ten years ago, Alibaba's technology might be very different today.

If our engineers, product managers, and operation service staff had already had the true partnership spirit, Alibaba's tomorrow would be even more different.

If the Doctor were just a programmer, then Alibaba's Chief Technology Officer might be the same as any other company's CTO.

If the conductor of a band needed to be proficient in various musical instruments to be a conductor, then many senior executives of the Alibaba Group would be laid off.

No one in Alibaba Group is born capable of being a CEO, CFO, or CPO (Chief People Officer). The great thing about Alibaba is that we can make someone who seems likely incapable of being any kind of "O" become a unique "O"!

Thanks, everyone, for your criticisms of the Doctor. It is these criticisms that make the Doctor more perfect.

Alibaba will turn a Doctor in Psychology into an excellent CTO, just as the United States could turn the actor Reagan into a president.

The Doctor brought about the achievements of today not only because of his talent and hard work but also because of the support and help he received from you and me. I hope everyone will continue to support the work of the Doctor.

August 15, 2012, written initially for
Alibaba's internal website, aliway.com

BEING ONLINE

Chapter 1

Confusion, Dislocation, and Entanglements in the Digital Era

Keywords five hundred thousand years ago are light *and* darkness, *keywords fifty years ago are* digital *and* analog, *and keywords today are* being online *and* being offline.

Confusion between Data and the Internet

Over the past several years, I encountered three seemingly unrelated events, neither major nor minor, that led me to contemplate seriously the internet, data, and computing.

On May 12, 2013, the Hangzhou Mobile Talk Club organized a cloud computing–themed salon at Yunqi Town in Hangzhou. The music platform Xiami's founder Xiaowei Wang spoke about the streaming music service, Kuaidi[1] founder Weixing Chen spoke about product design in the era of cloud computing, and website Duanqu[2] founder Qiangyu Wang shared his experience of doing start-ups in the cloud era. These gentlemen could be considered the old guard of

---

1   A start-up car service app like Uber.
2   A text message sharing service.

China's internet age, but they were early adopters of cloud computing. I too took the opportunity to express my own views about what was going on. Generally speaking, after having being thrown into a tizzy for some time by the rise of the mobile internet, everyone, including me, was considerably confused.

After the salon, Yuefei Pan, a reporter from the *Qianjiang Evening News*, compiled my thoughts about the position of the cloud with respect to the mobile internet, my reflections on big data, and my bleak outlook on innovation coming from app development and posted them on We-media under this provocative title: "Jian Wang: You Are Totally Wrong about Cloud Computing and Big Data."

Drones are a classic application of big data

I love peace, but the two best examples I can think of when I speak about cloud computing and big data both have to do with war.

One example comes from a talk given by the author of *The World Is Flat*. He went to the front lines of the war in Afghanistan and decided that West Point has to completely change the way it educates soldiers. He discovered that a low-ranking officer just needed a phone to command the movements of a fighter jet on an aircraft carrier. This could never have happened before in war. Previously, the ability of low-ranking officers to deploy war resources would have been extremely limited, but now they receive the same training as an army or division commander would have in the past. People often ask, "Can I change the world with the help of right infrastructure?" In the past, many major tasks could be carried out only by order of the president.

But today, thanks to cloud computing, an individual can do many more things than the president.

Another example is drones. Some people say that drones are the best use of big data in the world. Drones are a classic case of innovation. Wars could not have been fought like this before, but now, because of big data, everything has changed.

Cloud computing helps you do things you could never do before, and at scales that would have been previously impossible.

Our clients are the people who could best use data
When Aliyun [Chinese acronym for Alibaba Cloud] was founded, we said we are the first choice as a platform for data sharing. I myself believe that by using this market positioning we were ahead of our time. In the early days, we were focused on data, not information. This is a very important distinction. And when we mentioned data, we viewed ourselves as a platform. We believed that those best positioned to use data weren't us but rather our clients using the cloud.

Today, the most successful data company is Google. Google uses the Web data available to everyone around the world, and they relied on their processing ability to create the world's largest business. During Google's early years, other companies had access to the same data as Google, but no one else had their processing ability or vision of data.

When we were deciding how to position ourselves, we always maintained the basic concept that "data is just data." But there are definitely people smarter than us who could turn data into an important business.

The amount of data available today is far greater than before, so there are more business opportunities, and the opportunities are bigger than ever.

Before, no one knew where the data was, until Google launched its search engine. This seems simple today, but back then it was a very difficult task, because not many people had figured out how to turn the data resource that they had on hand into a profitable business. Otherwise, Microsoft and Yahoo! would not have outsourced their search engine services.

The term "big data" is a misnomer
I believe the term "big data" is a misnomer, because it doesn't reflect the essential nature of data.

Actually, big data has existed for a long time, it's just that the word "big" didn't yet have any meaning. The world's biggest data has nothing to do with the internet. The European Nuclear Research Organization (CERN)'s supercollider possibly had the greatest amount of data, more data than would be possible to completely process in an entire lifetime.

Data isn't meaningful simply because it's big. Today, data is meaningful when it goes online. This is the uniqueness of the internet. That everything can be online is far more reflective of the true nature of data. If the traffic data that Kuaidi used were not online, the app would be worthless. Taobao's[3] data has value because it is online. Data on a tape or paper has very limited usefulness.

In other words, the value of data becomes much greater and more obvious when it's online. In past presidential elections, Gallup used to have to randomly call

---

3   Taobao is a large e-commence website.

two thousand people and have them take a survey in order to get accurate predictions. Now we do not have to do that. An analysis of people's tweets can accurately predict who will be elected president. Furthermore, Gallup's results have difficulty having an immediate impact on society. Data, however, can have a quick effect. Similarly, ride-hailing software probably has a bigger impact on the drivers than even their companies simply because the data is online.

Sometimes executives at petroleum or geological exploration companies talk to me about their big data, but I don't actually believe that what they're talking about is indeed big data. They do have a considerable amount of data, but if it's not online, then it's meaningless.

Data is more important than features
I'm not a product expert. Anything that is wrong with Aliyun's products is entirely my fault. I hope everyone can forgive me for that. But after experiencing a lot of difficulties, I believe I've learned some things.

So far, I have yet to see a product that integrates data very well. In the pre-internet era, a product's selling point was its functions, but now the most valuable thing about an internet product is data. There are many apps that provide flight information, but the one I use most often is Umetrip. From a product perspective, I'm not sure how good it is, but its data is accurate and useful. It can tell you what happened to your incoming flight.

For software in the past, features were far more important than data, but I believe that in the future the opposite will be true. People in the internet industries understand this. I once joked with Jack Ma that we understood less about data than big-box electronics retailer Suning did about e-commerce. My point was that no one really knew where the opportunities were in this emerging industry.

A few days ago, I ran into Kai Ye, CEO of Play-crab (an online game company), and he said that after making online games for years, he finally realized data was the most important thing. Running back and forth copying data between computers is difficult and taxing. I would guess that he believes completely relying on intuition to make online games is not enough; data can help him do this better.

Cloud computing is a business based on trust

When I was thinking about cloud computing, I suddenly realized that it's a business that is about trust. Streaming music service Xiami and new media outlet Duanqu moved to Aliyun because of trust. China was the first country in the world to use paper money. Putting a stamp on a piece of paper and getting people to believe it was wealth was an extremely bold proposition. But now we Chinese have fallen behind: we don't have faith in credit cards and simply use them like debit cards. This is a huge problem.

If someone says he wants to create an internet business that generates 20 billion dollars in three years, and he wants to rely entirely on someone else's cloud computing services, he can believe in the cloud. That is courageous. I believe that without this type of courage, there can be no innovation. As of now, China has this type of opportunity.

Insurance is a very traditional industry. If you open an insurance company, you will have to have a physical office. No one would say that you can take care of insurance business without signing a paper letter of proxy. But, probably in the near future, China will have an insurance company that is the first without an office. You won't have to sign any piece of paper. Everything will be possible online.

Making an app is like planting in someone else's garden

I think mobile internet innovation can go beyond just developing apps. Developing apps is kind of like planting in someone else's garden. Apple and Android have already set the fences around their garden space, so it's okay if you plant some flowers, but it's just like buying a potted plant at the supermarket. If you really want to make something robust it will be very difficult. You need to go into the forest and clear your own land for something to thrive. Otherwise, if they want to take down your product, you have to take it down. This is a problem. Of course, some people believe the garden provides a great environment. But in China, there are many more revolutionary things we can do. Exactly what they are, we aren't certain yet.

There would be no Apple without Xerox

The greatest challenge we face in cloud computing is we still don't know exactly what cloud computing is. This is an emerging industry. We had only learned about it in books, and it took everyone's hard work and cooperation to make it a reality. I think there should be an internet museum. Today everyone knows the internet, but they don't know its history, which makes it hard to innovate.

Many people probably don't know that almost all mouses are made in China, most are sold in China, and much of the technology for their construction is also in China. Very rarely do people ask why it's called a mouse, and even fewer people know that the first mouse was made of wood.

Today everyone can see the beautiful iPhone, its operating system, and the excellent App Store. But most people don't know where this device came from. A documentary about the 1980s in the United States features an interview with Steve Jobs in which he says if he had never been forced out of Apple, he wouldn't have realized his later accomplishments. If Jobs hadn't been pushed out, he wouldn't have gone on to found NeXT, which wouldn't have been bought by Apple, and therefore we wouldn't have the Apple we do today. From this, we can see it is very hard to plan out the serendipity required for innovation.

I do believe that Aliyun can succeed, but it will not do so without going through tough times. There is a rumor on the internet that after Microsoft came out with Windows, Apple sued them for copying the Macintosh OS. Gates and Jobs eventually decided not to make a big deal of it, concluding that they both had stolen ideas from Xerox PARC. Xerox PARC didn't commercialize this technology well, but they made a major contribution. Without Xerox PARC, there would be no Apple. That is to say, start-ups all make products that might fail, which are nonetheless valuable.

It's not about what data you have, it's about what service you provide

Today, sensors in a cell phone aren't very different from the sensors on a cruise missile. They can generate a lot of data, but it isn't all that valuable. Data collection is not a space full of innovation. If you can think it up, it's doable. The area with innovation potential is what you do with all the data you have. This is a topic no one has really figured out yet.

This is a chicken-and-egg problem. If no one does the foundational work well now, then the later innovations will not come. People have been talking about location-based services for a while. But as of today, even though we know where users are, no one has developed good services around this. I think this is the greatest challenge moving forward.

The ambiguity of "cloud computing"

Taking cloud computing as the present-day equivalent of electricity in the industrial age, data is like a Ford assembly line. Just as there is no way there could have been widespread industrialization without electricity, without cloud computing we wouldn't have big data. I believe there is ambiguity surrounding the term "cloud computing." When we say "cloud computing," we're really talking about computing in the cloud, big data, and storage in the cloud together. Considering only one piece of the puzzle leads people to see only the big data of Ford's assembly line and forget about the cloud computing that makes it possible. For that reason, I distinguish between the visible front end and the invisible back end. I don't believe it when people say they can do big data without the cloud. Without the cloud, you're just a small workshop.

Don't think about using data to improve existing services

Banks today don't use big data on their own because they lack the right computing infrastructure. They rely on IBM machines to process data. If you calculated the costs of processing data, you'd discover the value produced by processing data cannot be canceled out by the cost of processing itself.

Data's great strength is that you can maximize its value with minimal costs.

This isn't to say that any place that has data will definitely have a big data business. The best data services at Alibaba are in finance, but it isn't the same as banking. Alibaba Finance is doing microfinance, which just happens to be a field that banks don't work in. Banks don't have this data, and carrying out a credit check is expensive, so they don't do microloans. Alibaba Finance gives out loans of a few hundred RMB [around $50] to countless people every day. There was a client who borrowed only 1 RMB, and he wrote a special letter of gratitude saying that no one had ever been willing to lend him 1 RMB in his entire life, and after Alibaba gave him the 1 RMB loan, he felt like he had been respected for the first time.

Don't think about using data to improve existing services. That isn't what data is meant for. It should be used to do things that couldn't be done before. Amazon does product recommendations better than anyone else in the world, but that's still the early days of big data. If Amazon is thinking only about how to use big data to further improve its recommendations every day, then I think they will definitely have no future for their business.

Yuefei Pan later wrote in the lead-in to the above article, "That day, another We-media author tweeted out several of Jian Wang's quotes using a public WeChat account. It resulted in losing the most followers the author had ever lost in a day and sparked considerable discussion among the account's followers. This is the impact of Jian Wang, who is one of Alibaba's most controversial leaders—his fans love him, but his detractors hate him." Even though I hadn't particularly prepared for this salon, the dramatic effect of my remarks was a big surprise for me.

Dislocation of the Cloud and Computing

In January 2013, I was interviewed in depth by Jiashi Li, a reporter from *China Electronics News*, a major IT media company. The conclusion I gave in that interview was that cloud computing was no longer a theory or vision, it was a viable business.

Jiashi and I first met when I was still working at Microsoft Research Asia in Beijing. Later, after I'd moved back to Hangzhou to join Alibaba, she found out that I'd sparked controversy with my work launching and running Alibaba's cloud business and YunOS (a mobile operating system). Some people said I was someone really passionate about technology who had a noble soul, while others said I wasn't the right person to be running product R&D and a business. In October 2013, the night before the Aliyun Developer Conference, she interviewed me regarding the cloud and computing.

This interview was scheduled after a meeting Jian
Wang had with an investment company, so it was
pushed back to ten o'clock in the evening, but despite
his bloodshot eyes, Wang spoke with vigor and animation about the cloud. He discussed the cloud,
China's basic software R&D, the thinking about so-
called internet thinking [a buzzword], Singles' Day
sales [an e-commerce sale day that occurs on November 11, similar to Black Friday in the United States],
and 12306 [a train ticketing website in China, which
Aliyun helped to support]. In the past, I have interviewed Wang and shared his thoughts with other
tech executives, who have all found him to be very
insightful. Currently, China's IT industry is undergoing a transition while the cloud and the internet
are causing shocks to many industry sectors. There
are many paths that the promotion of the cloud and
the internet could take. Certainly, there are also considerable misunderstandings. I hope Wang's outlook
will inspire others in China's IT industry as it undergoes this transition.

**Most companies are approaching the cloud in the
wrong way**
The cloud is currently divided into three types by the
industry: public, private, and hybrid. IBM, Microsoft,
VMware, and other IT giants are pushing these three
types of clouds depending on their client's needs.

But Jian Wang believes, "The private cloud isn't
really the cloud at all." Cloud computing's essence is
service. If you can't scale the computing resources and
make them accessible to a large group and you can't
really provide them as a service, then that isn't really

something that can be called cloud computing. For that reason, he believes IBM and the other traditional tech giants providing the private cloud have at most added some technological sophistication and some new features to their current product offerings without actually achieving a fundamental technology shift. Instead, they're repackaging it and selling it as an upgraded IT solution.

Wang believes that the development of cloud computing requires an ecosystem, but when several so-called cloud computing alliances sent him invitations, he turned them down. After reviewing these groups' members and their charters, he decided they were superficial and worthless, and that they had essentially adopted the IBM and Microsoft mindset. Instead, Wang set up his own cloud computing alliance in Hangzhou, called the Yunqi alliance [Yunqi means "where the cloud dwells"]. This alliance came about not because it had been Wang's idea all along but rather because, in the course of Aliyun's operations, its users and developers encountered problems and it became clear there was a need for such an organization. So, he decided to bring together Aliyun's disparate developers, users, and partners to create a real platform where everyone could work together to solve their problems.

During the course of several interviews, Wang expressed his concerns for the way China was pushing cloud computing. Everyone knows that cloud computing takes a service model, "but when you really do it, you try to sell equipment to the users." This core business operation model hasn't yet been changed. Currently, China's many cloud computing service providers, including telecom companies and local governments, are all building massive data centers. They've bought tens of thousands of servers, storage space, and networks, but "99 percent of them aren't really cloud computing." So, Aliyun made Apsara, which is something akin to a cloud operating system. By cooperating with the data centers, Apsara can "cloudify" their computing resources and turn them into a service that people outside the companies can use.

Wang says there are only one and a half companies that Aliyun regards as its role models. Amazon truly realizes the service potential of the cloud. The other half company is Google, which has solved the problem of scaling up but hasn't completely done a good job with services.

Cloud security is the topic that is most discussed and the greatest source of concern. Wang says, "In reality, the cloud is safer than traditional storage. It's like putting money in the bank or hiding it under your mattress. Which one is safer? Obviously, it's the bank. Using the cloud only requires overcoming a psychological barrier."

Only through every enterprise embracing the internet does the cloud have value

In many interviews, Wang has emphasized repeatedly that if the attitude of users and the ecosystem don't change to embrace the internet, the cloud will be meaningless. Only when users and the whole ecosystem adopt the internet will the true value of the cloud become obvious.

He believes that many users are still employing the old IT mindset. For example, in order to deal with the spike of traffic on Singles' Day, online retailers need to decide how many servers to buy; this will cost them at least a few thousands RMB. Not to mention the additional cost of maintenance in the future. Subsequently, when they come to Aliyun, Aliyun helps them use cloud storage services to handle Singles' Day for only a few RMB.

I asked Wang, even if they were able to handle the massive volume of transactions on Singles' Day, how could they possibly deal with the problem of train ticketing on the 12306 website during Chinese New Year?[4] He said, if you compared the volume of actual transactions for these two projects, Single's Day is much bigger than 12306, and the conclusion is obvious.

As Aliyun's list of clients continued to grow, someone once asked Wang when big finance clients like ICBC, one of the world's largest banks, would start using Aliyun. Wang answered: "You should ask ICBC this question. If they believe they are an internet company, then by all means, they should use Aliyun services. At the very least, the vast majority of their non-core services could be put on the cloud."

Now that sectors from finance and retail to manufacturing are facing the challenge of the internet, it is almost as if no industry is unaffected. When promoting the transition of traditional enterprises into internet enterprises, there are many options and models.

---

4 Actually, the 12306 train ticketing website started using Aliyun services in 2015.

But which will result in the most successful transition to internet-based companies?

According to Wang, many people now view the internet merely as a channel for existing services and products rather than seeing it as a change in attitude and business models. Wang said, "To be considered a true internet enterprise, the company has to have these two features: first, they're using cloud computing; second, they have to use data to optimize the services they provide through the cloud."

Some people may believe Wang's opinions are too extreme, but I agree that his vision expresses a true understanding of how the internet is transforming business. If we don't experience a sea change in our core attitudes, we won't be able to increase efficiency and effectiveness. Only a great change can bring great results.

Wang said, "Using the cloud" is in essence "using the internet." Most companies' IT and services rely on the internet, which is challenging traditional companies in ways they're unable to handle. If your company has nothing to do with the internet right now, it would be difficult to imagine it still being around in ten years.

The cloud is the opportunity for system software
Jian Wang believes that only with the cloud can China's IT industry transform itself.

China's system software has never been able to take off because there has been no opportunity to interact with users. Wang said, "You can't make a good product with the door closed. The development of a product is always the result of interaction with the users." Take databases, for example. Databases weren't built in a day. They were developed over time through constant conversation with users and continuous growth. Now that traditional databases have turned into a large-scale and matured market, we no longer have the opportunity to interact with users that allows us to grow together with them, and we are now facing insurmountable obstacles.

"I often say that demand is competitiveness," Wang said. "China's cloud computing could someday overtake America's because China has much greater demand for it." He compared the situation to Taobao [the number one e-commerce website]. Taobao is able to have such explosive growth because China has such large market demand. Whereas in the United States, Walmart was already well-known and widespread. This is one of the reasons the development of e-commerce was more difficult in the Unites States at the beginning. Demand can tell you where the problems are. In Wang's opinion, competitiveness isn't learned from books, it's formed by the demands of users. Problems are solved as they're encountered, and this process cultivates competitiveness. That was the case when Aliyun was developing Apsara, the centerpiece of the Aliyun technology architecture. They ran into many problems and were frequently criticized. Wang said they were able to grow and succeed at Aliyun only because they worked hard and addressed the problems.

In his opinion, system software has a fresh opportunity today because of the cloud. We're now leaving the software era and entering the cloud computing era, which means the whole structure of the software system is undergoing a sea change. Development of software based on the cloud gives China another chance. But Wang looks at the companies that received funding for projects related to core electronic components, high-end chips, and system software and believes there still exists a misunderstanding. Many firms, including large ones, are still operating under the attitude of "I have what you have," in which companies compete to simply replace existing products that are already on the market. Lacking viability in the marketplace, these will at best become vanity projects.

Currently, Alibaba is promoting the de-IOE process, which refers to removing the reliance on IBM minicomputers, the Oracle database, and EMC storage and involves moving from commercial software to open-source, and then eventually to "proprietary tech + cloud." Wang said that Chinese users' demand for the cloud is like dry kindling waiting for a match, and he hopes that not only system software but apps and the entire IT ecosystem will be able to more clearly see the opportunities inherent in the cloud, innovate based on the cloud and the internet, and as soon as possible revolutionize China's IT sector, which has always been playing catch-up with other countries.

In 2015, IBM shifted from the "smart planet" to cognitive computing and began talking about Watson, an artificial intelligence initiative. In January 2021, the *Wall Street Journal* reported that IBM is considering selling its Watson business. Who knows where their focus will move in the future? Microsoft had already started to firmly embrace the cloud with the strategy of "Cloud-first," while Google was still struggling to catch up. Silicon Valley is wondering whether investing large amounts of money and human resources into app development is the best way to show America's innovative capacity. Obviously, people are looking for the best way forward.

I thought at the time that my discussion with Jiashi Li wasn't well organized. But when the article appeared in *China Electronic News*, it came with the headline, "Alibaba CTO Wang Jian's Outlook on the Cloud and the Internet." That really surprised me.

Entanglement of Being Offline and Being Online

I often use Skype to make calls. One day, I got frustrated because there are messages Skype can never send out. A friend told me, "If you're offline, my messages will never be sent out to you. It isn't that convenient to use."

Indeed, the way Skype has configured its settings, if you send a message to me and I'm offline, it won't be received. If you send me something when I'm offline and then later you go offline, the message still won't go through until we're both online at the same time. A post on the social networking site douban.com (an online community of sharing after reading books, listening to music, and watching movies) complained about this issue with Skype. According to the site, it took two people a whole month for a message to be received due to this setting. Today, even after many rounds of improvements, the file-transfer experience is still not consistent across different versions of Skype due to architecture and business reasons.

This example clearly demonstrates the offline mentality. In Skype's mind, you have an offline life that is separate from your online life. You'll only go on Skype to be online when you're in your office or home in front of your computer. If you're in the subway, out jogging, or eating, you'll be offline. Skype provides service only when you're actually using the app; just as when you make a phone call, both people need to be present for it to work. But this is viewing the internet like a telephone system. I believe that if Skype can't overcome this binary view of the world as being offline and being online, it won't survive the competition in the mobile internet era.

Skype provides free internet phone calling services around the world so you no longer need to be limited by telephone companies or national boundaries to make quality voice communication. "When I downloaded Skype, I realized the traditional telecommunications era was over," said Michael Powell, former chairman of the Federal Communications Commission. Powell may have seen the value of telecommunications, but he didn't necessarily see the future of the internet. Otherwise, there would be no WeChat today.

Like Skype, Tencent's QQ messaging service is also stuck between the worlds of being offline and being online. QQ is a national app in China. It was launched in 1999 and now has more than 650 million users, making it the most popular messaging app in the world.

QQ introduced the term "online" to Chinese users and made it familiar. When you wanted to chat with people or send your coworker a document, you would have to click on the QQ icon on your desktop, log in with your username and password, and finally, set your status as online. When you hear someone in a third- or fourth-tier city say, "Go on QQ," what they mean is "go online." Being online has become another term for using the internet.

At the beginning of 2011, WeChat was launched. Another Tencent messaging product, it is a free communications app that relies on smartphones and can be used across operating systems and various communication carriers. Unlike the smartphone version of QQ, WeChat has no eye-catching online setting—it was designed in the belief that users are online twenty-four hours a day. Only 433 days after its launch, WeChat had already garnered 100 million users, a feat that took QQ years to accomplish.

QQ quickly responded first by just translating their PC-based product into a phone-based one. Then, in the May 2013 phone app update, they had learned from their WeChat colleagues, and not only did the design look more like WeChat but, more importantly, the online setting had disappeared. The tech media reacted strangely with headlines like "QQ launches knock-off version of itself," expressing their dismay at the new update.

But users were even more outraged. The update had been live for only four days, but 38,000 users in the App Store had already weighed in, with 90 percent giving a negative review and some calling it "secondhand WeChat." Most users were upset with one specific change: the removal of the online setting.

I believe the change that QQ made was necessary and not a simple matter of copying. The mobile internet has made the online setting quaint and a thing of the past. Only by removing that setting can the internet penetrate into every corner of one's life. "Always being online" is now the default. You believe you're always online, but also that there is no single action that puts you online. Only when that's the case will the demands of online society be met.

Ultimately, QQ will have to delete the offline function, something it can't yet do only because of its users' habit, cultivated over the course of more than a decade. WeChat has no such problem because it never had an offline setting. What QQ needs to overcome is not WeChat but something much harder—the offline world. QQ eventually relented and in another so-called "update" restored the offline function. Today, users of QQ are still happy with this offline function.

QQ's regressive update and Skype's stubbornness regarding being offline vs. being online aren't simply a struggle within particular products. Rather, I see in such struggles a sign of the future.

In the year 2015, there were many discussions in China about the future of the internet. More often than not, these discussions ignored important details that have a bearing on our future path. This again surprised me.

These surprising things made me feel as though a new world was upon us.

Four hundred thousand years ago, Peking Man was using fire, which gave humans warmth and light. With fire, a dark cave could become a bright place. The world was divided into darkness and light. Light was the root of survival, while darkness was the source of fear. In the 1980s, new discoveries in Africa showed that humans had been using fire much earlier than previously thought—now it was known that humans had control of fire 1.42 million years ago. Light had driven out the darkness, so we were living in a world of brightness.

Starting in the 1950s, digital technology arose and digital computers began replacing analog computers. We were moving from the electricity age into the information age, with computers transforming the structure of society and its values. Today, every word we write is digital, as is every book we read and every picture we take. Digital has become part of modern humanity's DNA. Digital has swallowed analog, and now it's clear we live in the digital age.

Today, the internet continues to develop at high speed. Big data, cloud computing, the mobile internet, artificial intelligence, wearables, robots, virtual reality, the internet of things, blockchain, and many other new terms are appearing throughout various media outlets, forums, and expos. Every day of the internet era seems to give birth to a new technological innovation. Science and technology are like a wild horse roaming free amid the concrete jungle. It is hard to tell the direction the horse is going in, but it affects everyone.

In my opinion, in the background there is an invisible and powerful hand in control. It's the term we're both familiar and unfamiliar with: "being online." This is the most important keyword of our era.

Every day I meet new innovators. As someone who's on the front lines of technological advancement, I think constantly about the origin of these developments, where they come from, and what is powering the changes. The only way to grasp the changing path of the internet is to understand the unwritten rules it follows. Being online is both common sense and yet also at the core of my way of thinking about the internet. Which makes me want to discuss this oft-ignored piece of common sense, brash though this may seem. The nature of common sense is that it is truly basic, foundational, and long-lasting.

Being online is the common sense of the coming new world. Examining it, observing it, thinking about it will help eliminate the confusion it engenders.

Every civilizational inflection point has its own unique keywords, and today the one we face is "being online." The online world isn't a change from an old one; rather it is a New World, an undiscovered continent. In the remainder of this book, I will use a new way of thinking to approach the world we're living in today—this world that has been transformed from being offline to being online. I'll share with you my understanding and what I've learned about being online in the hope of dispelling confusion, dislocation, and entanglements.

Chapter 2

The Evolution from Being Digital to Being Online

Bits were made from atoms during the information revolution before the internet. The rise of the internet is putting bits online. Being online is an unknown continent to be explored, just as North America once was.

Many years ago, I made several trips to Xinjiang and Tibet and realized that we're getting closer and closer to the world of being online.

Kucha is a small county nestled in the middle of the Tian Shan mountains at the northern edge of the Tarim Basin. The county is also referred to as Qiuci and its airport as Qiuci Airport. When I drove south through the city along National Highway 3012, there was nothing green to be seen anywhere around me. Everywhere was brown rock. The whole area was devoid of any sign of human life. There I felt as lonely as a man on Mars. I felt like I had returned to a primordial state. There is nothing but wilderness upon wilderness.

The road traversed many hills, making the drive difficult and forcing me at times to go as slow as three miles per hour. While I was driving at a snail's pace down the road, I suddenly noticed power lines, winding wires carrying with them the possibility of any number of uses of electricity. That was the only thing in the area that had anything to do with humans. I was taken aback to see those wires in such a desolate place.

Today, as soon as people gather together and set up places to live, the first thing they do is make sure electricity can reach them. Where there is electricity, there are people. And where there are people, there certainly is electricity. My reaction at the time was "electricity is truly amazing." The power line is the symbol of the electrification era; it is the connection to industrial civilization; it brings us closer and closer to civilization.

After visiting Xinjiang, I traveled to an even more remote location—Tibet. Signs of human life are sparse on the road to Tibet, where the remoteness is so great it isn't even connected to the grid and there are no power lines to be seen in most of the area. But on a roadside devoid of electric lines, I saw a surprising sign: CELL PHONE RECEPTION IN 5 KILOMETERS. That meant that just five kilometers down the road was a telecommunications base station, and a cell phone could access the internet. I could be online and reconnect with the world. Thanks to solar power, online civilization can reach places without power lines. This was yet another reminder for me that internet connectivity is more important than electric lines. Even in the absence of the grid, you can access the online world of the internet. This is the future of world civilization: being online.

If you go back thirty years, the words "being online" were on the lips of the cutting-edge technologists of the day. But now, everyone has heard "being online" so many times they're practically sick of it.

In 1978, Bill von Meister founded a company called The Source, which provided an early online service, Bulletin Board System (BBS). At the time, users could get online with

300-to-1,200-baud-rate dial-up modems. The Source defined itself not as hardware or software but as a company that could connect your personal computer to anywhere in the world. Von Meister later founded the Control Video Corporation (CVC), which provided online game services for the home video game Atari 2600. Not long ago, I found a promotional disc for America Online (AOL) in an old book I have. AOL's predecessor was CVC. AOL was the first to bring a wide range of services online—news, stocks, weather, retail, photo sharing, search, email, real-time communications, and so on. And in 1998 when it announced it was buying out Netscape, less than 1 percent of American internet users had access to broadband.

Although AOL and Time-Warner's merger in 2000 wasn't successful, *USA Today* in 2007 still ranked the founding of AOL fourth on a list of the twenty-five most important internet-related events in the preceding twenty-five years, after the World Wide Web, email, and the graphical user interface (GUI). Numbers 5 and 6 were broadband and Google, respectively. You can still feel the impact of AOL on some in society today through various events. In 2010, military intelligence analyst Chelsea Manning, then known as Bradley Manning, leaked hundreds of thousands of American military and government classified documents to WikiLeaks, including the July 2007 footage of an American Apache helicopter firing on Iraqi civilians, revealing the truth about the grave situation in Iraq and Afghanistan and sparking protest around the world. Manning first used AOL Instant Messenger to publicly assert their possession of such classified materials.

This is not just the impact of AOL. It is the impact of being online.

China also used to have several companies that included the term "online" in their name. For example, there were the websites Zhongguancun[1] Online and Zhejiang Online, which now seem old hat. But whether these names are outmoded or not, online is now always at our side and has penetrated into every aspect of our lives. It is no longer possible to avoid the issue of entering the online world.

Being Online Is the True Nature of the Internet

Before the internet era, there were a billion mouse clicks a day, but that was considered a minor event that couldn't be translated into value for traditional software companies. After the internet put clicks online, clicks became a source of wealth for Silicon Valley.

I first started thinking deeply about the idea of being online when I was working for Microsoft. As the world's most successful software company, Microsoft would put each of their products, from Windows to Office, through rounds of usability testing. Simply put, in order to make their products more user-friendly, Microsoft would observe representative clients using their products to carry out typical tasks in a usability lab or in the field. They would then be able to see if their products met their users' demands and how they could be designed to be easier to use.

To use a term from the internet era, "usability testing" was meant to create a better "user experience," or UX. Back when software was still installed using disks, Microsoft was already very focused on UX, and every product had a specialized team working on it. The most common way of conducting usability testing brought average yet typical users into

---

1   Zhongguancun is considered the Silicon Valley of China.

specially designed labs where they would be watched through a one-way mirror, recorded, and given questionnaires and surveys. Sometimes a team would even go into users' offices to observe their most authentic behavior. Then, after extensive data analysis, they would discover product flaws, correct them, and ultimately release the product.

This was once common practice in the software industry and the only feasible way to carry out UX in the offline world. Microsoft had thousands of engineers who worked exclusively on these tests, which were the only way to learn about users' interactions with their products.

But this method had limits as to how many users it could evaluate and could only ever cover a small minority. Further, even though the usability teams could go to users' offices, they couldn't very well follow them home and see how they used their products there. Microsoft eventually came to their senses, realizing that the old way of product testing could cover only tens of thousands of people, while their products sold hundreds of millions, perhaps over a billion, copies. They understood that even though they had been selling software for decades, they really didn't know how their products were used in the real world. This may sound strange, but it's true.

Before the internet, Microsoft's great regret was that they had no way to know how most users actually used their products in real-life situations. In the internet era, a mindset formed from this traditional practice could prove fatal, if it meant someone else was closer to your clients than you were.

In the early twenty-first century, Microsoft did something very innovative. When Windows 95 had become widespread, so too had the use of the internet. When Microsoft designed Office 2003, they built in technology that would track usage data and send it back to Microsoft over the internet. This was the User Experience Improvement Program and was known internally at Microsoft by the acronym SQM, for Software Quality Measurement. At this time, Microsoft engineers could write code to track any particular problem they wanted to understand, and the software would automatically collect the relevant data and send it to Microsoft. What made this method stand out was that it made data feedback a regular part of the software development process. MSN's product team used SQM to analyze system performance and understand user behavior, and Windows Vista became the first operating system to have been developed with SQM involved from the very beginning.

According to the blog by Jensen Harris on msdn.com, after Office 2003 was released, over 1.3 billion user sessions were collected, with each session recording SQM for a specific period of time. Over the last ninety days of the data collection, Word alone recorded over 350 million clicks. Microsoft was able, for the first time in software development history, to discover that the five most common commands are paste, copy, save, delete, and bold. These five clicks together accounted for 32 percent of all commands given in Word 2003. That many users providing data from real life would have been inconceivable before the internet.

I believe Microsoft did a "half-internet" job. Today, when you install their software, you're asked whether you want to join the user experience improvement program; in other words, whether you give permission to have your usage data collected and sent to Microsoft. This is a big step forward, because even though the company is using traditional software, it's still taking advantage of the internet to improve its software experience offerings. You are getting many thousand or tens of thousands times as much data as before, and it's definitely coming from users' real working environments rather than being gathered by lab observers. At the time, this was an incredible accomplishment indeed.

We have already gotten used to this type of process today, thanks to Microsoft's innovation.

When I first came across this, I thought it was strange because I didn't understand how users would be willing to have their every mouse click sent back to Microsoft. Harris's blog answered my concern.

When I looked at the volume of the data, I was shocked. Can you guess how much usage data is gathered? Back when I was working there, back-end server systems weren't powerful enough to store all the data. An algorithm had to be designed to throw away a significant amount of data intelligently. It's amazing how people all around the world are content to send in their data.

But if we look at this SQM process more closely, we can see that it is overly complicated and bloated. Microsoft first needs to design which usage data it wants to collect and then write extremely complicated code in order to collect the data online. This is a difficult process and has a cost.

Unfortunately, the cost of collecting data for a traditional software company is very high. A click when using their software is fed back to the software itself when it carries out the task you wanted it to, but that click isn't necessarily sent back to the software company. In order to collect usage data, the software company must first conduct research and add a layer of code.

Fortunately, internet companies are online from the day they're formed. In the past, software was a purely offline product that was forced to be online with the emergence of the internet. This was difficult, but the process was not complete. However, internet companies represent a major shift from software companies.

For an internet company, the cost of collecting data is virtually zero. Websites and web pages naturally need clicks from users to work, thus the precipitate of clicks naturally becomes data. This is why everyone pays so much attention to cookies (stored data to personalize Web services). Think about it. If you click on something on a web page, this data is automatically transmitted back to the company; otherwise the company would have no way of showing you what was clicked on. The only thing they have to decide is for how long should this data be stored. Dealing with data is like an instinctual response for an internet company. It is just like when you instantly move your hand away from a cup of boiling hot water.

Google is a classic example of an internet company. Born in the dorms of Stanford University, the company has been giving the world pleasant surprises year after year since 1998. Today it is considered the world's most used search engine. Google's success has long been seen as the result of processing information and optimizing advertising. But behind these successes lies something much more important, something that everyone has overlooked—a countless number of clicks.

Before the internet, the billions of mouse clicks that happened every day didn't matter very much. Most software companies didn't pay attention to them at all. Only Microsoft adopted SQM. No one had yet turned clicks into gold, because at the time all clicks were offline and simply dissolved in your local PC.

The internet brought these clicks online. Clicking a hyperlink could allow a machine to know what you wanted more than you did. When Google took offline advertising that was done in newspapers and turned it into AdWords, clicks suddenly changed. Today a click can be worth thousands of dollars, transforming clicks into the source of wealth for Google. From then on, clicks had a new value, a new form, a new space. From clicks, the most influential internet company was born.

Today it's not just clicks—our whole lives have gone online.

Your computer can go online. Every click you make, every purchase you make online, every website you visit, every key you tap are all online.

Your cell phone can go online. Cell phones have GPS, meaning your location can be online. Cell phones also have motion sensors, which allow for your every move to be recorded online.

Your eyes can go online. In 2012, Google Glass was released, meaning everything you see could be online and be the part of the world. In other words, the whole world will look totally different in an online society. There was an article at the time that said Glass was requiring society to come up with a whole new set of rules and laws in our daily social life. There was ample logic to that argument.

Your car can go online. Tesla, BMW, and Apple are all trying to add smart technology to cars. The next time I go to Tibet, I believe I won't have to look for a sign saying when there is a cell phone signal because the car will likely be connected to the internet the whole time with the recent deployment of Starlink by SpaceX, which means I could be in contact with a colleague on the other side of the world to discuss business the whole trip.

Your air conditioner can be online. The air conditioner has sensors that can check temperature and airflow, which become data that you can view remotely.

Eventually, even your sleep will be able to go online. As long as you have a wearable device, how deep your sleep is and how much time you spend in bed vs. how long you are asleep can be tracked and measured. I could provide even more examples. The internet's penetration into our lives can make mundane things around you brand new.

In the future, twenty-four hours a day, whether you are sleeping or awake, you will be online every second. Being online is becoming the new essence of this era.

Atoms Become Bits, Bits Go Online

Bits were made from atoms during the information revolution before the internet. The rise of the internet is putting bits online. From a technological and philosophical perspective, being online first requires us to be linked, connected. In turn, connection is the result of being online. This kind of connectivity will penetrate all of society, changing industries and lives.

Mouse, clicks, World Wide Web, and online

Being online and connectivity couldn't exist without several seemingly minor, often-overlooked technological advancements: the mouse, hyperlinks, and networked computers.

These technologies are intertwined with the life of one person: Douglas Engelbart.

In 1945, as World War II was drawing to a close, Engelbart was a radar technician for the US Navy in the Philippines. While in the Red Cross library on a small island, he first read Vannevar Bush's classic essay, "As We May Think." It was inspiring and laid the intellectual foundation for Engelbart's later work. This essay also had a profound impact on the World Wide Web (WWW) and information technology in general. In certain respects, its influence outstrips that of the Manhattan Project, which Bush himself had helped organize.

After graduating from UC Berkeley in 1957, Engelbart entered the Stanford Research Institute and founded the independent research group known as ARC. In 1963, he invented the mouse, which was little more than a wooden box with a red button on the right corner and two wheels that produced a positioning signal. Because there was an electric cord coming out of the box that looked like a mouse's tail, the device quickly picked up the nickname of "mouse." Engelbart in his 1968 patent application assigned it the lofty sounding name of "display system X-Y position indicator." In 1968, Engelbart gave the first demonstration of the mouse in San Francisco, and he received the patent in 1970. But at the time, no one knew what this small device was worth, and Engelbart didn't make a fortune from it. Years later, Stanford sold the technology to Apple Computer for $40,000.

In 1972, Xerox Palo Alto Research Center successfully developed a graphical interface microcomputer (an early PC) that they called the Alto. Some of the researchers at the facility had previously studied under Engelbart, and so they decided to attach the mouse to the microcomputer, which made the machine extremely easy to operate. In 1983, Apple emulated Xerox's design and added an improved mouse to the Lisa microcomputer, a move that made the mouse well-known. As the internet began to spread around the world, the mouse was in hot pursuit. Although most people believe the mouse to be only twenty or so years old, in reality it has already lasted five decades.

Looking back on the last half century of technological development, this small invention has had an outsized impact. The mouse is the second thing you plug into your computer, following the keyboard, and it has made our ability to interact with computers much more convenient. Little wonder it was named by the Institute of Electrical and Electronics Engineers (IEEE) one of the most important events in the fifty years since the birth of the computer. Yet most people know nothing about such details, just as most don't realize that almost every mouse in the world is made in China. These two seemingly unimportant facts reveal the large gap between innovation and playing catch-up. A wooden box invented in America in the 1960s would become a major product of China's manufacturing sector in the 1990s.

The mouse is only one of the many contributions that Engelbart made to the development of the internet and computing.

On December 9, 1968, at the US Fall Joint Computer Conference (the precursor to the American National Computer Conference) in San Francisco, Engelbart displayed the world's first mouse. In a ninety-minute presentation, Engelbart became the first person to show the mouse expanding and shrinking the size of a window and controlling the text on the screen (a black dot on the screen later became the cursor on every computer screen). Engelbart was connecting to colleagues from about thirty miles away in Menlo Park via a modem. This was also the first remote videoconference, for Engelbart to speak with them. It was the first time clicking with a mouse on underscored text would take you to another page. In other words, the first mouse click in front of a public audience also took place at the conference. It isn't a coincidence that the hyperlink and the mouse were displayed together for the first time by the same person.

This presentation was termed, by journalist Steven Levy in his 1994 book *Insanely Great*, "the mother of all demos." The impact was massive. It directly links to the development of the Xerox PC the Alto in the 1970s, the Apple Macintosh in the 1980s, and the Microsoft Windows operating system in the 1990s. Stanford University held the thirtieth and fortieth anniversary conferences for this presentation in 1998 and 2008 respectively, probably the only conferences in the world to celebrate the anniversary of a presentation. It's worth noting that the Stanford Research Institute was one of the four initial access points of the Advanced Research Projects Agency Network (ARPANET), and Engelbart was one of the first thirteen employees there. ARPANET later evolved into what we now call the internet.

The mouse was the most important revolution for the information age. Now the mouse is being replaced by touch-pads and touchscreens, but we still need to click. If there were no mouse, there could have been no hyperlinks to jump to different pages, and everything that came out of those innovations would have been impossible.

At the beginning, everyone was surprised by the ability to jump from one page to another simply by clicking on text. Over time, we've become accustomed to being able to click on anything and that action taking us somewhere else. If you Google "pu-erh tea," for example, and it doesn't immediately unveil a page showing the price of the tea and the best method of steeping, you'll think something is wrong with the website. Even my neighbor's four-year-old child knows this from watching cartoons on his iPad.

"Online" originally meant you were linked up to your computer. If you use a printer, you'll discover something very funny: the printer's settings are displayed as online and offline. Online means that it is connected to a computer, and offline means it is not.

Actually, people have talked about being online for a long time. There were theories about being "forever online," "everywhere online," and "anywhere online." But the one thing they didn't address clearly was that being online required the internet.

When I was writing this book, I connected my laptop to a twenty-seven-inch monitor and put my draft on the monitor so I could use my laptop screen to look up more information on the internet. Although the large screen was physically connected to my laptop, it wasn't online. The internet knew my browser was up on my laptop but had no idea that there was a large monitor sitting there. The browser was online, but the monitor was connected only physically. Therefore, I believe being online and being connected are two deeply intertwined concepts, but they have clear differences.

When people first discussed the internet, they were actually talking only about physical connectivity, like fiber optic cables and other infrastructure. It's easy to confuse physical connectivity with being online. But they're not equivalent. In Chinese, we say *hulianwang* (the Chinese translation of internet literally means "interconnected network"), but that is the internet plus the World Wide Web in English. Here the internet refers to the physical connection, and the World Wide Web means online content. Because the World Wide Web accounted for more than 90 percent of the traditional internet's content in the early days, the two concepts are often confused with each other. It should be said that the true use of the internet is to provide a platform that can make previously unconnected digital content go online. What is the most important thing on the Web? It is yet another thing that people have overlooked: the ability of text to be clicked on, redirecting the user elsewhere, i.e., the hyperlink. The internet connects devices, but hyperlinks connect content. That is revolutionary. But it's only the first step of the internet. The greater meaning of the internet is that it can bring even more things online. The internet of things that people frequently talk about refers exactly to the internet's ability to put online more real-world items.

Connectivity and being online may often be confused, but in reality they are two separate things.

When you build a road, pedestrians, cars, and bicycles are connected from point A to point B. But the way you connect is hugely important. If you're walking, the road quality doesn't matter all that much because a small path and a concrete road will get you there at the same speed. The number of people you can connect this way is limited, and you cannot connect up with people all around the world.

But if you're driving a car, things are different. The road quality becomes very important. At this point, you may start thinking about living in the suburbs, but without roads and cars, the suburbs wouldn't even exist. At the same time, you have more opportunities to meet with people from all around—the region, the country, and the world.

Today you have the opportunity to access all the things in the world if you go online. Facebook can connect you with people from every corner of the globe. In that sense, connectivity is indeed being online. Connecting to the internet is like linking up to all the servers in the world. Having a cell phone that can go online means you are connected to everyone on earth. Previously, we had access to a computer; today, we are connecting to all the servers and all the people in the world. This example again shows that we shouldn't assume connecting is the same as being online. These are two totally different things.

We exist originally in the offline world. Equipment, goods, and people are all offline. Traditional computing technology turned physical offline things into digital offline things. Internet technology took the offline online, thereby bringing tremendous changes to humanity that could someday outstrip the impact of humans first using fire.

There are many online roleplaying games. You need servers and computing power in order to support a virtual person in a game, and computing power comes from electricity consumption, which in turn produces carbon emissions. According to researchers' calculations, supporting a virtual person online results in as much carbon emission as raising a real human. The line between the offline and online worlds is getting blurred, as offline gradually disappears and the world evolves into a completely online society. We are finally "being online."

It is difficult to define in one sentence what being online is, and it is too early to define it at this point in the discussion. Just as the word "digital" has two different translations in Chinese, the same word has two different translations for different contexts. Each translation carries a different bias. Being online connects all the servers in the world and you with everyone in different parts of the world.

In the past, the most difficult thing was getting data online. Now wearable technologies are putting data online that was previously uncollectable. The essence of big data is being online, and being online is bidirectional.

Clicks being online resulted in the success of Google. This can be understood as the source of all inputs (links of search results) going online. Your behavior, which is the source of all outputs (your clicks on links of search results), is going online.

Early Web portals could only display content; they couldn't receive feedback from users who read it. You could only view content from newspapers that had been put online. This was a one-way street. Gradually, websites began to prepare smart recommendations for you based on your reading habits, browsing data, and interaction information. Such recommendations were closer to the true application of big data, but it was still not enough. There is a news app called Jinri Toutiao, or Today's Headlines, developed by ByteDance. The company claims that it takes five seconds to calculate your interests and has intelligent recommendations based on that. This is based on the concept of the two-way nature of being online. You input your interests and hobbies, and the site will present you with related news and information, creating a "news engine without an editor." The app produces none of its own news, but within two years, it had 120 million users, 40 million of whom were active monthly users. In June 2014, it was valued at over $500 million, greater than the value of most traditional media outlets. This was due to its use of online applications and machine learning. TikTok is also from ByteDance. The technology to make targeted recommendations may partly explain why TikTok is popular in United States.

Data must be online, and the output process should be online too.

YouTube, Hulu, and online data

YouTube is the world's largest video-sharing platform. It is different from other video websites because its sources are entirely online and its distribution is entirely online. The majority of YouTube uploaders are individuals, but some media companies like CBS, the BBC, Vevo (a music service site), and other organizations that have partnership programs with YouTube upload their own content onto the site.

Hulu is an American video site that was launched by NBC and Fox in March 2007. Many people compare Hulu with YouTube, saying that the only difference is the former has copyrights for their videos and the latter doesn't.

In my opinion, Hulu, which has intellectual property rights to its content, is simply taking something that is offline and putting it online. YouTube, however, supposedly lacks IP rights, but it uses online videos as its source of content. If there were no YouTube, videos would only ever be made in the traditional way—first filming, then saving to a computer and editing, then posting online. This is certainly less valuable to society.

On April 29, 2009, a pregnant English woman felt pain in her stomach and thought she was about to give birth. Her husband rushed to YouTube and searched for videos on how to assist in a home birth. Four hours later, the woman did start to give birth, and her husband successfully delivered the baby based on what he had learned on YouTube.

If you compare Hulu versus YouTube, which one do you think has greater value to society? Of course, it's YouTube. Hulu is only putting traditional media online.

Internet companies often discuss user experience, but I believe that term has been overused. Every internet company cares about user experience (UX). But strangely, the internet is the only industry where UX is all that's worth talking about. According to that logic, as long as the software is well designed — that is, the UX is done well — then you won't have any other problems. In reality, other industries do care about UX while they also need to solve other challenges as well.

A very interesting example of this comes from the history of the American automotive industry. The competition between carmakers was intense, which led to rampant copycatting. There was one company that decided to add an extra coffee cup holder in its design, a move that ended up becoming the main reason consumers decided to buy the model. For car buyers, comparable cars were the same in most aspects, so having an extra cup holder would lead to a better experience, making that particular car worth buying.

The internet's online user experience is unique in its essence, inasmuch as it differs from the offline experience. This is key. The internet is different from traditional industries not because it provides a good user experience but rather because the experience of being online is a leap beyond traditional experience. Therefore, it is important not to put the cart before the horse.

UX should not be a kind of "accurate nonsense." Looking back, when the internet first appeared, how bad was the experience? How bad was the first mouse's user experience? UX is not the unique strength of the internet; it is something that must change with the times. Today, it's fair to say that the mouse and the moveable cursor were the best experience possible at the time.

People can do many things using different tools, but sometimes using tools will result in a decrease in the ability to control. For example, if there are two circles on the screen, using your fingers allows you accurately move back and forth between circles. This is because you're using your fingers directly instead of some device. But imagine if you used a chopstick to click on things. Your accuracy and speed would drop dramatically. In other words, using your finger to click on things is more accurate and faster than using something else. When there is an object that can enhance your abilities, you will gain something but also lose something.

But how are the mouse and the cursor connected? The mouse is one of the few physical add-ons: when you move it in your hand, the cursor's movement accuracy is on par with using your finger. For that reason, using a mouse isn't considered troublesome or awkward. That's what's amazing about the mouse's design. The mouse follows Fitts's Law (named after Paul Fitts, who was a psychologist at Ohio State University), which says the time required to move to a target area is a function of the ratio between the distance to the target and the width of the target. That ratio won't drop simply because a tool is added; so, even though you're using a mouse, it doesn't reduce accuracy compared with your bare hands. (The significance of Fitts's law is best described in the book *User Friendly: How the Hidden Rules of Design Are Changing the Way We Live, Work, and Play* by Cliff Kuang and Robert Fabricant, which was published in 2019.)

There are many things that could replace the mouse, but none is nearly as natural. More specifically, even though the mouse is something that isn't part of your body, it essentially works as an extension of your hand. To this day, I have yet to see a device that can compare with the mouse in this respect.

Thanks to the mouse, we have clicks. Before the touch screen, the internet economy was basically the click economy. Before the internet, the hardest thing to do was collect clicks on a computer and turn them into big data. But everything changed after the internet. Today's wearables are putting mobile computing online as they put previously unrecordable data online. Even though people are endlessly talking about wearable computing, it is essentially no different from mobile computing. It's just mobile computing that's worn on your body. Currently, wearables have solved sensor problems but not computing problems. For example, bracelets like Fitbit can measure your heart rate, thereby using sensors to solve a digitization problem. Digitizing data is extremely important, but it's just the beginning. As soon as the data goes online, then another revolution occurs.

Today's wearables are revolutionary because they are connected to the internet.

In August 17, 2013, the AllThingsD tech website published an article headlined NBA REFS WEARING GOOGLE GLASS?, based on an interview with Houston Rockets general manager Daryl Morey. Morey was asked, "Do you think Google Glass could ever be integrated into live games, for example by having the officials wear them and letting the TV audience see what they saw?" Morey's response was simple "This is a great idea!" It's worth noting that Morey has a tech background. Later, the media asked him to confirm whether or not his comment was a joke, and Morey actually went a step further, saying that NBA referees wearing Google Glass was "absolutely an interesting idea" that was worth a try.

This is a classic example. If you thought of wearables only as sensors, you're wrong. As soon as the referees start wearing Google Glass, whatever the refs see will suddenly be online. At that point, the impact of wearables on society will be completely different. Everyone will know what a ref knows, so if he blows his whistle when looking at another part of the court, the fans will immediately be aware.

If not only every ref but also every player wears Google Glass, then everything the players see will be online. That would be an extremely interesting view. The winner wouldn't be Google Glass but rather "being online."

Today, virtual reality, augmented reality, and mixed reality are hot topics. Magic Leap, once a company with buzz, claimed to be working on cinematic reality. People get all entangled in what reality in general is. In Miami once, I joked with Rony Abovitz, founder and then CEO of Magic Leap, that it's all about online reality. Reality in the past wasn't online; today reality is being online with the assistance of virtual, augmented, mixed, or cinematic technology.

The Map: An Evolution from Analog to Digital to Online

Being online should be measured and discussed in units of centuries.

Microsoft MapPoint and Google Maps

Maps are perhaps the world's most frequently used type of data.

In China, the map had already appeared in the myth of the Chinese Emperor Yu and the legend of him controlling a huge flood in the Yellow River Basin dating back to 4,000 years ago.

According to the *Zuo zhuan* (Zuo tradition), an ancient history compiled around 400 BCE, it was during this time in the Xia dynasty (2070 BCE) that nine images were recorded on nine large cauldrons. Each image showed one of the nine regions of Xia with their particular topographic features—rivers, mountains, grasslands, forests, even animals. Later, because the Qin invaded the Zhou, the cauldrons were lost but were perpetuated by the ancient book of *The Classic of Mountains and Seas*, which emerged from this type of imagery.

In the West, the oldest extant map consists of nearly 4,500-year-old ceramic fragments from Babylon. A piece about the size of a person's palm shows the mountains in the north from which the Tigris and the Euphrates originate, and the swamplands in the south through which these two rivers flow, with ancient Babylon in the middle.

Maps have always held an important place in the development of human civilization.

In 228 BCE., the Qin kingdom sacked the Zhao capital at Handan, destroying the Zhao kingdom. The Yan kingdom worried it would be next, so Crown Prince Dan sent an assassin named Jing Ke to kill the Qin king, who would later become Qin Shi Huang, the first emperor of all of China. As a way to ingratiate himself with the Qin king, Jing Ke brought with him a map of a region of Yan, which the king viewed personally in the presence of Jing Ke, giving the assassin an opportunity to carry out his mission, which ultimately failed. From this, you can see how important maps were to ancient people. There is a saying from the *Zhou Li* (an ancient Chinese book of Confucianism) that goes, "The officer in charge of land shall report with a land map and evaluate land quality and conditions." This is perhaps the oldest reference to maps in Chinese writings. The *Zhou Li* also says, "Disputes over land are to be resolved by maps." Here, the economic value of maps in ancient times is clear.

Maps are a measure of a society's knowledge. In the Early Middle Ages, a period also known as the Dark Ages in the West because of the monopolistic role played by religion and the loss of knowledge from antiquity, the view that the world was round was under attack. At that time, maps were no longer representations of geographical knowledge but rather served as illustrations for works of theology. This type of map almost always drew the Earth as flat, without latitude and longitude lines, and they were never to scale. It was not until around the year 1000, when an enlightened way of thinking arose, that cartography and geography appeared as disciplines and began to impact other fields.

Links between the Old and New Worlds were based on maps. In 1538, Dutch cartographer Gerardus Mercator made the first map of the whole world. This map used a view of the Earth from the vantage point of the North Pole, and there was a southern part of the map that was labeled "unknown southern continent." This map was made toward the middle of the Age of Discovery, shortly after Ferdinand Magellan circumnavigated the globe. It was the summation of all human knowledge of the world's geography.

In the bits era, satellite imaging, remote sensing, GPS, and navigation systems have resulted in the rise of digital maps.

I remember a digital map that came on disks and could display far more information than was available on a normal map on paper. You could use any scale to zoom in and out, thereby making data that was legible only to experts available in a straightforward image. Digital map disks used to be an important business for Microsoft, which sold an astronomical number of copies every year.

But the map disks were quickly surpassed, because in 2005 Google launched Google Maps, an online map. From then on, no one ever thought about maps on disks again, and the companies that made those digital maps were soon forgotten. Few people know today that at one point Microsoft sold MapPoint on Amazon for $299.95.

Disks lost out not simply because Google Maps was better or the data was more accurate. Rather it was because Google put map services and data online. From that moment on, maps were no longer the same thing they once were.

In 2008 came news that caused quite a stir. Google had launched its own satellite to take photos for its map, meaning Google was now gathering its own mapping information. For many people, the first time they saw satellite images of Earth was on Google Maps.

Actually, that news item was inaccurate. The satellite was launched by DigitalGlobe. Google Maps used high-resolution images from two geospatial image companies, Digital-Globe and GeoEye. These two companies merged in 2012 and continued to provide a considerable amount of data, including satellite images, to Google, which simply made this data available online. Interestingly, DigitalGlobe spent only $453 million to buy out GeoEye, meaning that simply having data is not valuable in and of itself.

In that case, where does the value and impact of putting that data online come from? You can tell from a 2009 *Time* magazine story that listed the top ten discoveries on Google Maps. This top ten list included possible evidence of a submerged landmass that was thought to be the lost continent of Atlantis, missiles fired during military exercises, and an airplane graveyard full of mothballed B-52s and stealth bombers. These online high-resolution images made it possible for anyone with a connection to the internet to search the Earth with a fine-tooth comb.

Many things were discovered by online communities of hobby-ists that had never been seen before. It was this type of commu-nity power that rediscovered the value of maps. Traditional research institutes had no way of reviewing all 230,000 square miles of images produced by DigitalGlobe every day.

Nokia tried to learn from Google's methods. In 2008, Nokia spent $8.1 billion on Chicago-based Navteq. Their logic was simple: a cell phone had to have maps. This was not a bad idea, but still Nokia failed. Nokia didn't put a map online; instead, they just installed a map on cell phones. All they did was take an offline map and put it on a phone, then they promoted cell-based guid-ance systems, including walking guides. After all, with only 2G, it was very hard for maps to go online, so it seemed to make sense just to focus on guidance systems. Nokia didn't understand that the major trend of this era is being online, and guidance systems can be offline but they comprise only a small part of the usage of map services. In November 2012, when Nokia launched a new brand called HERE, Navteq, once valued at the $8.1 billion, had already disappeared. Similarly, if I hadn't brought up HERE here, I bet no one would remember it once existed.

What Google did was take someone else's data and put it online. This data produced even more data after it was used. It wasn't a case of new maps that defeated old maps; it was online that defeated offline. This was a sign of the changing times.

It is difficult to maximize offline data's value, but putting it online can. Offline data isn't likely to be the most competi-tive, but online it can be. And while it's not likely that offline data will have a huge impact, online data does.

Being online is the cause, and transformation is the effect. Any minor insignificant thing that goes online can have a massive transformative impact.

When I was in elementary school, there was a stand posi-tioned beneath the traffic lights. A traffic officer would sit in the stand to make sure traffic flowed smoothly. What the traffic officer did was to manually change the traffic lights based on his visual observation of traffic flow from another direction. Over time, the technology improved, and this role was automated. Using closed-circuit television cameras, traffic flow controls, cameras that catch red light violators, speed sensors, and other digitized methods, the function of the traffic stand could be automated, and now the city is being driven by data. I once told a mayor, "I don't like the term 'smart cities.' I prefer 'data city,' because whether something is smart or not is too subjective." But this isn't enough. The data hasn't been put online completely. If city planners want to do traffic planning, they usually make a new design by looking at the data of the previous year. The data used for that analysis is used once. It isn't available to improve traffic after that because it isn't actively online anytime, anywhere.

In the past, when you bought something in a store, everyone knew that the way things were displayed was based on many considerations and the arrangement would affect sales. At the time, it was very important for people with experience to determine the placement of products. But Taobao has more than 1 billion products. Where could you find someone with the skill to place that many products? Only data can do that for you. Actually, Taobao's appearance changes every day, and its visitors are presented with different pages because it uses the data from the previous day to rearrange the display of products on the page. Each of your clicks affects what you see on your pages the next time you log in. On Taobao, if there is a special situation, we can temporarily change the prices and arrangement of products. It is very easy to do because we have ready access to the data. But as Taobao grows and its number of users increases, merchants need to adapt in real time. This need drives up the demand for data. It requires keeping the data online, ensuring it flows freely at all times. Actually, Amazon treats its online product placement in a very similar way.

This is the difference between something that is driven by online data and something that simply relies on data. It's not that I don't believe digitization is important; digitization is the foundation of being online. The most successfully digitized things in the world have to do with multimedia. In the past, the best digitized product was newspapers printed digitally. The Founder Typesetting System (Founder is a leading digital system provider for newspaper publishing in China) was very popular, because it used laser printing technology for newspaper composition instead of traditional lead blocks. This was a digitizing process. When it came about, the whole traditional printing industry disappeared. However, the process still ended by producing a print newspaper. Though everything in the process was digitized, the end product was a physical replica, a paper newspaper. This is a step backward. With respect to the end result, these two different processes for printing newspapers weren't actually all that different. But, at least at the beginning of the process, the newspaper is digitized. Now that there is the world of being online, the next step for the news media itself is not hard to imagine. Before the laser printing process, it would have been impossible to make the leap to being online. Even though they're both related to computing, digitization and being online are two different realms. It's the difference between Founder and Google.

Kodak, Sony, digital photos, and digital music
Another product that relied on digitization to revolutionize an industry and depended on being online to generate value is the digital camera.

In December 1975, at Kodak headquarters in Rochester, New York, Steven Sasson became the first person in history to hold a digital camera and take a picture with it. Sasson was among the first group of electrical engineers Kodak hired in the early 1970s, a break from the company's practice of almost exclusively hiring chemical and mechanical engineers. Sasson's dream was to "Take pictures without film, look at pictures without paper." In 1978, he was awarded the patent for the digital camera. The year the patent expired, 1995, three new digital camera models from Casio, Canon, and Logitech brought the price down by around $1,000. Mr. Sasson was awarded the National Medal of Technology and Innovation at a White House ceremony in 2009.

Today it's hard to find a non-digital camera. Even if you're a hobbyist and actually still have film left, you don't use it, but view film photography more as work for art collections and not as a regular consumer product. In this respect, there no longer exist film cameras on the market, or they exist in a niche or resale market. Kodak once made their own digital cameras and added Wi-Fi capability, which counts as an improvement. But that was not a revolution, since it only made it easier to download the photos onto a computer.

The new revolutionary product in cameras is the digital camera on smartphones, which can connect to the internet at any time. An online camera completely changed the whole nature of photography. Now, when you go out to eat, it doesn't matter whether it is a first grader or a forty-five-year-old, you'll see all sorts of people take out their phones to take pictures of their food and post them on social media. Online cameras have sparked normal people's interest in creating and sharing. That is why a site like Instagram could become so popular around the world.

In January 2012, the 131-year-old Kodak company filed for bankruptcy. From the first digital camera to bankruptcy took less than forty years. Today Sasson's vision has been realized, but Kodak practically does not exist. This shows the difference between Kodak and Instagram.

Music went through a similar process.

From cassettes to CDs and then to MP3s, it has become much easier for us to access music. This followed the process of transitioning from physical to digital and finally to being online. Many technologies appear and, before they have even had time to find their niche, quickly disappear. For example, the portable CD player hadn't yet been perfected when it was already being replaced. In November 2012, Japan's Dr. Serkan Toto tweeted: "Sony is now selling Japanese music on iTunes. I thought this day would never come." Eight years before, because of the competition between the iPod and the Sony Walkman, Sony Music refused to provide any Japanese artists to iTunes. That Sony finally released its catalog not only reveals the difference between the Walkman and iTunes but also demonstrates the advantage of being online over just digital. That is the difference between Apple and Sony.

In the history of human development, many things have evolved but very few things have persisted. From a utilitarian perspective, this is similar to the principle of the survival of the fittest. Things like glasses, watches, and necklaces haven't been replaced, because they have been adopted like natural extensions of the human body. I believe being online is a similar concept. The evolutionary cycle of being online should be measured in centuries, and there needs to be enough time for it to evolve to be part of everyday life. Certainly, we ourselves need a suitable amount of time to properly understand it.

Being Online Is a Brand-New Continent

In 2006, the famous mathematician Shing-tung Yau became embroiled in controversy surrounding the Poincaré conjecture. Henri Poincaré was a French mathematician whose greatest accomplishment was in proposing several pioneering math problems. The Poincaré conjecture proposed in 1904 is but one of many: "Every simply connected, closed 3-manifold is homeomorphic to the 3-sphere." In May 2000, the Clay Mathematics Institute listed the Poincaré conjecture as one of its seven Millennium Prize Problems. According to the prize rules, anyone who published a proof of any of the seven problems in a mathematics journal and allowed that proof to be reviewed for two years would win $1 million.

Perelman and his new continent

In early June 2006, Xinhua News Agency reported that Harvard professor and Fields Medal recipient Shing-tung Yau had announced, at the math research center of the Chinese Academy of Sciences, that, based on the work of American and Russian scientists, Xiping Zhu, a professor at Sun Yat-sen University, and Huai-Dong Cao, an associate professor at Tsinghua University, had proven the Poincaré conjecture. Their proof had been published in a special June issue of the US-based *Asian Journal of Mathematics* and took up more than three hundred pages. The article was titled: "A Complete Proof of the Poincaré and Geometrization Conjectures — Application of the Hamilton–Perelman Theory of the Ricci Flow."

In late June, at an international conference in Beijing, Yau gave a keynote address on the Poincaré conjecture and for the first time in public announced that the conjecture had been completely proven. There were more than a thousand people in the audience, including the renowned physicist Stephen Hawking.

In late July, the *Wall Street Journal* ran an article by Sharon Begley in which she quoted 1962 Fields Prize winner John Milnor saying that Yau's "completely proven" statement had thrown a monkey wrench into the process of sorting out who actually had proved the conjecture.

On August 28, 2006, *The New Yorker* published an article by Columbia University professor of journalism Sylvia Nasar and David Gruber about this Poincaré conjecture controversy, titled "Manifold Destiny." Nasar was the author of *A Beautiful Mind*, the biography of mathematics genius John Nash, which was later made into a movie. This added gravitas to their article.

On September 1, 2006, Weiyue Ding, who is a member of Chinese Academy of Sciences and head of the Peking University Mathematics Research Institute, wrote a post on his Sina blog entitled "Poincaré's confusion." He wrote: "Prof. Shing-tung Yau is a widely recognized global authority on geometry. There is no way he did not know that many mathematicians around the world believed Russian mathematician Grigori Perelman had already solved the Poincaré conjecture. His announcement was at odds with a great many of his colleagues' beliefs."

However, Perelman had never met the requirement of the Clay Institute to publish his proof in a peer-reviewed journal. He spent eight years researching this hundred-year-old math problem and then, between November 2002 and July 2003, published three drafts of papers on arXiv.org, a website for math and physics theses. This site is maintained by Cornell University's library and looks like a simple and crude website. Importantly, its articles are not peer reviewed.

In May 2006, Bruce Kleiner and John Lott, originally from the University of Michigan, published an article, "Notes on Perelman's Papers," on arXiv. The same month, a panel of nine mathematicians agreed to award the Fields Medal, the mathematics equivalent of the Nobel Prize, to Perelman for his solution to the Poincaré conjecture.

In June 2006, International Mathematical Union chair John Ball personally traveled to Saint Petersburg, Russia, to invite Perelman to accept the Fields Medal, but Perelman rejected his offer.

Later that month, Nasar and Gruber made a special visit to Saint Petersburg in preparation for writing "Manifold Destiny." They reported in the article that Perelman said "He is not a math hero," and "If everyone knows the proof is correct, then there is no need to have any other means of recognition."

In July, John Morgan and Gang Tian published a 473-page draft on arXiv titled "The Ricci Flow and the Poincaré Conjecture." In an interview of May 2006, Morgan said that Perelman deserved the Fields Medal because he had solved the Poincaré conjecture.

In August, at the International Mathematicians Union Conference held in Madrid, Perelman was awarded the Fields Medal but didn't attend the conference. He was the first and is the only person to ever have rejected a Fields Medal.

In September, *Notices of the AMS* (the journal of the American Mathematical Society) published a long review article by deputy editor Allyn Jackson about the whole episode, in which he quoted James Carlson of the Clay Mathematics Institute as saying, "According to the rules of the award, the recipient is not necessarily going to be the first person who publishes the result in a journal. . . . In reality, Perelman chose to post his result on arXiv rather than submit it to a publicly circulated journal." This not only affirmed Perelman's contribution; it also was an affirmation of the internet.

This outcome is something that could never have happened before. Because of academia's peer-review principle, scientists might not have given credence to Perelman's result, since it was neither peer-reviewed nor published in an academic journal. But because Perelman posted his article online, the whole world could know how he had done his proof, and it didn't need to be peer-reviewed any longer. Additionally, all work done subsequently on the conjecture would just turn out to be taking Perelman's work a step further.

In December 2006, Xiping Zhu and Huaidong Cao published a revised version of their article in the *Asian Journal of Mathematics* in which they concurred with Morgan and Tian in their appraisal of Perelman's successful solution to the conjecture.

In March 2010, Grigori Perelman was given the Millennium Prize Problem award. However, he didn't go to Paris to attend the award ceremony in June. In July, Perelman formally rejected the $1 million prize, saying the money meant nothing to him.

According to its Wikipedia entry, when it was launched, arXiv helped spark the academic open access movement, and now it is commonplace for mathematicians and scientists to post articles there before sending them to professional journals.

I don't understand mathematics at all, but I believe that, through this controversy, we can see the power of the internet.

New continent of knowledge

The scale of goods and things has risen exponentially because of being online. The scale of thinking is also being transformed as computing goes online.

The *Encyclopedia Britannica* was once the world's largest book, and it was the most famous and most authoritative encyclopedia. One of the world's three major encyclopedias, it was born in the eighteenth century during the Scottish Enlightenment. The first edition began being compiled in 1768, a process that took three years. In 1771, a three-volume set was published. From 1788 to 1797, the third edition had grown massively to eighteen volumes, plus two appendices, which brought the page count up to more than 16,000 pages. The average person couldn't read all of those pages in their lifetime.

Today, the scale of the *Encyclopedia Britannica* is not a big deal. Wikipedia has 3.8 million entries, thirty-eight times that of *Britannica*.

More importantly, Wikipedia is a free and open Web-based encyclopedia that is the fifth most popular website in the world. Entries are edited and written by people in every country around the world. The site's use of wiki technology means any person from any background and any country can edit any entry. An offline encyclopedia cannot possibly compete with that. The root of this change is in the rise of being online.

One piece of common sense that is closely related to your life but has turned out to be untrue is the idea of the eight-hour workday. Now you no longer know when you're working and when you're living. The line between the two is blurring, thanks to online products like email, collaborative working, and instant messaging services.

We're doing more than what we thought before. When cameras still used film, it would have been impressive for a normal consumer to shoot dozens of rolls. Only a photography hobbyist who traveled frequently would be able to do that. But with the advent of digital cameras, the number of photos taken exploded exponentially in an instant. Taking 100,000 photos in a year is now normal. After cell phone cameras became commonplace and photography therefore went online, those 100,000 photos on your computer—out of which you'd occasionally select a few to print out and put on your nightstand—also grew exponentially. By putting them online instantly, you can now share them with thousands, maybe even hundreds of thousands of people at once.

A company that uses cloud computing properly can, even if it has only one employee, access the computing power of a ten-thousand-person corporation.

Thanks to being online, costs, access to users, and speed have all changed. Speed creates value. This is at the core of internet innovation. In the book *The Facebook Effect*, there is a case study showing how one person can impact the whole world. A Colombian civil engineer named Oscar Morales was fed up with the horrors brought on his compatriots by drug traffickers and the government's weak response. At that time, the FARC, a narco-terrorist group, had taken more than seven hundred people hostage, shocking the nation and giving people the feeling that their whole country had been taken hostage. Morales logged onto Facebook and discovered that there were no FARC-related groups, activities, or posts, so he formed the first anti-FARC Facebook group, which had the slogan, "No kidnapping, no lies, no massacres, no FARC." He named the group, "A Million Voices Against FARC."

As more and more people joined the group, he called for a nationwide anti-FARC protest. The result was amazing. Not only did people all around Colombia participate, but members of the diaspora in Miami, LA, Buenos Aires, Madrid, Paris, and other cities also joined in. The result was an unprecedented global protest launched online. On February 4, 2008, an estimated 10 million Colombians in nearly a hundred cities marched in protest, with nearly 2 million people in cities around the world joining them.

No one would have guessed that this massive protest movement was launched by a frustrated young man sitting in his bedroom, posting an impassioned article on Facebook. Overnight, it shook the world. That is the speed of being online.

Speed is the foundation of innovation. This foundation does not serve to create new ideas that can never become a reality. Rather, it helps you turn new ideas into a service that can be beneficial to 10 million, thereby creating value.

Someone once said that the speed at which a new technology is adopted by 50 million people reveals the extent of its penetration. Electricity took fifty to sixty years to reach 50 million people. But today it could take just two years for a new technology to garner 100 million users. That is the power of being online.

The *Encyclopedia Britannica* has already become unimportant to most people. When the journals *Nature* and *Science* also lose their importance, just as the steam engine is no longer as important to us today, when researchers post the data from their experiments directly online instead of sharing the results only by publishing papers, society will take a step forward, and the world will become more beautiful.

Being online is a new continent. We need to explore it more deeply.

Chapter 3

The Evolution from Information to Data

Being online is one aspect of the world. If there were no data online, then we would not be able to discover many new things. This is not an era of data becoming big data. it's about information being turned into data.

Data Is More Valuable than Information

The Ford assembly line was a marvelous achievement that epitomized the Industrial Revolution.

Prior to Ford, carmakers would put their vehicles together by hand. This meant high personnel costs and considerable time consumed, making automobiles very expensive and a symbol of wealth at the time. Henry Ford was inspired by the way pigs were slaughtered and packaged, and he developed the first automobile assembly line in the world. It not only lowered workers' burdens, it also drastically lowered car prices. Therefore, Ford became known as the man who put the world on wheels.

But Ford's success was predicated on the already mature steel-manufacturing and electric-power-generation industries.

In the online world, cloud computing is the equivalent of power generation, and cloud storage is like the steel industry. Big data, meanwhile, is like Ford's assembly line.

Without steel and electricity, there would be no way to support large-scale industrial production. Similarly, without cloud computing, there would be no cloud storage, and without either, there would be no big data.

Data gets its life from being online

But before talking about big data, I have to say that I believe we should reconsider the term "big data."

In December 2012, I attended the 2012 Cloud World Conference in Beijing's Yizhuang cloud computing base. I took part in the Cloud Evening Talk forum, which centered on big data. When it came time for me to speak, I was very direct: "Actually, I believe the term 'big data' is a misnomer. It doesn't reflect the true nature of the phenomenon. The difference between the data of today and the big data of yesterday is that today's data is online." At the time, I knew one of the coauthors of the book *Big Data*, Viktor Mayer-Schönberger, was in the audience.

"Big data" creates misunderstanding because it makes people think the important aspect is the bigness. Is the scale of big data large? Of course, but that doesn't need to be overemphasized.

In Western culture, suddenly calling data that has always existed "big" might not be a big deal. But in China, calling something "big" is likely to cause habitual confusion. There is a book entitled *Connected* in which authors Nicholas Christakis and James Fowler discuss the impact of social networks, but when it was translated into Chinese, the book's title became *Big Connections*.

Overemphasizing big is hyperbolic; anyone given to sensationalism could then focus exclusively on that which is purely big. This tendency is deceptive because big is not enough. You still have to look at the truth of the matter. Big data has been around for a long time. The biggest data in the world probably has nothing to do with the internet. The supercollider run by the European Organization for Nuclear Research (CERN) produces so much data per collision that the world's most powerful computer wouldn't be able to process all of it in a human lifetime. That's the world's biggest data. In the lab, a 100-million-pixel camera takes 40 million photographs of atoms colliding per second. This produces 1,000 terabytes of data to be analyzed.

According to information furnished by Tim Bell, who is responsible for the CERN IT Operating System and Infrastructure, the supercollider records around 35,000 terabytes of data per year, and as the machine is upgraded, that number will continue to rise. Bell says 45,000 disk drives are used as backup storage devices.

These drives aren't online and therefore in reality hold dead data. This might be why there is no story about the supercollider in Viktor Mayer-Schönberger's book *Big Data*. The supercollider experiments gave birth to large-scale data, but that data is still a long way from being called "big."

Mayer-Schönberger also wrote the book entitled *Delete*. The subtitle to the Chinese edition seems very strange: "Making Choices in the Age of Big Data." The book sets forth a very logical premise: the most amazing thing about people is not what we remember but rather what ends up sticking in our memories. It's like the process of sculpting. You remove the parts you don't want, and what is left stands out. The book explores this idea the whole way through, but the Chinese subtitle attempts to tie its thesis to big data, which turns out to be rather awkward. This also happens with the book *Predictive Analytics*, whose title was translated as *Big*

Data Prediction in its Chinese edition. Here, the problem is more than obvious.

The term "big data" has already become overused. It's the buzzword of the era, so everything needs to be related to it. I'm concerned that if things continue like this, everyone will simply try to make their data big, and there will be chaos. Perhaps it will be like this: as soon as you mention big data, everyone rushes to build the biggest data center and buy the most computers. A company with a hundred thousand computers will definitely be better than a company with fifty thousand. Media reports will join in the fun, reporting on which company has the bigger data centers. If nothing changes, this kind of chaotic situation will almost certainly arise. In order to get people back on the right track, I would rather stubbornly repeat the main point of this chapter: "big" is not what matters; being online is.

Today's data has not gotten "big." The truly interesting thing is that today's data is online, taking advantage of the most important characteristic of the internet. Online data is definitely big. But data must be online for that bigness to become valuable.

Everything online better reflects the essence of this phenomenon than the notion of "bigness." From a more absolute perspective, digitized data is not in itself big data.

For data to generate value, it needs to first be able to be shared.

How should it be shared? By taking a hard drive from East China and carrying it with you to South China? Sharing is an expensive process. Without the cost-saving mechanism of being online, large-scale sharing is impossible. I am not certain about what will come later, but being online is definitely the development trend of the future.

Data must flow freely in order for it to have value. Data that is put online flows freely over the internet. This is the essential difference between the information society and the online society. If you have to use the word "big," then what is getting bigger isn't just scale, it's the space over which the data flows and the time frame over which it is accessible and being collected. Only in this way will the value of data also grow bigger.

Data Is a Strategic Resource

When discussing big data, everyone treats it as a gold mine, thinking it valuable, and they want to dig it up. But I believe big data's power lies in its ability to turn sand into silicon.

Silicon is an essential component of semiconductors and integrated circuits. Silicon Valley at its very beginning was focused on making semiconductors. Its electronics processing industry became famous the world over, and now it is the leading high-tech zone. The primary component of the sand beneath our feet is silicon. It is hard for some people to accept the fact that a key part of computer processors and an indispensable engine of the information age is all around us.

Intel, founded in 1968 and now the world's largest semiconductor manufacturer, is one of the best companies at making sand into silicon. The microprocessors it produces helped revolutionize computing and the internet, and changed the world.

Gold might be a strategic resource in a country. But that resource alone can't make the country globally competitive. True competitiveness is turning what everyone else has into something of value. Many people talk about how the data they have is important, rare, and unique. This is misleading in a certain sense. I'm not saying it's wrong but that it comes from a gold-prospecting mentality: having gold as a resource is the same as wealth. As for silicon, having silicon as a resource doesn't mean much. Only a company like Intel can extract the true value from silicon through its technologies. In the era of AI, the company that can best utilize silicon prevails. Nvidia is one of them.

I have said that clicks are not valuable, but putting clicks online is. This same logic can be applied to silicon and big data.

On the internet, the Web is there, just like sand. There is a lot of it and it's free; no one can monopolize all the content of the internet, and everyone can use it. These internet "grains of sand" are not valuable, at least they weren't before Google made them searchable. Search turned them into the silicon of data. By the same token, those people who brag about how much unique data they own will have difficulty turning data into their core competitiveness.

"Data is core competitiveness." I understand this in two ways: first, data as resources is core competitiveness, and second, the ability to process data is core competitiveness.

When everyone is on the same platform, then data can flow freely across that platform. When everyone is on the cloud, the costs of circulating and operating will drop significantly. With cloud infrastructure, data can be exchanged to create greater value. But without infrastructure, that is impossible. Walmart's success is due to sacrificing personalized products for low cost. This requires large-scale manufacturing and good infrastructure, including the electric grid. In the future, manufacturing will rely on the cloud and big data as well. A new economic structure will be formed in this way. That's the internet industry as well.

In the past, it was easy to measure a country's level of economic development by seeing how much electricity it used. When looking at the Earth from outer space, you can see that North America is very bright, because there are two very developed countries there and a lot of electricity is used. If you look at Africa, you see that most of it is dark, because very little electricity is used there. In the future, there will be a new standard for measuring a country's competitiveness — how much computing power it uses. This is the impact of the cloud as infrastructure.

China is one of the world's biggest contributors of global data and also one of the biggest users of servers. But as of now, there is no company that can say it can process all of the information on the internet, because that would require an incredibly large amount of processing and computing power. Having computers doesn't mean you have computing power. In the future, computing power will be something that is traded between countries. Currently, China is big in computer servers but certainly not in computing power. China must combine computing in the cloud with data to generate value in the digital era. Only in that way can China develop core competitiveness in the digital economy.

If the last thirty years was a computer and software era, then the next thirty years will be a time of computing and data.

Cloud computing is not mysterious at all. It is the data that seems mysterious, but data has also given us an opportunity. I'm very confident that in my lifetime we will see the combination of the world's two greatest computing devices: the internet and the human brain. In other words, the internet will become part of the brain. That is why the internet will be at the core of technologies in the next thirty to fifty years. When the brain becomes integrated with the internet, or when the internet is a part of our brain, we will be discussing the next revolutionary technology beyond the technology of the internet. This is my understanding of cloud computing.

The Telescope, the Microscope, and Radar

Using data to solve existing problems is not the ultimate value of data. Data can help you explore problems that you haven't yet discovered and then solve them. I believe the value of data can be best demonstrated by comparison to three important inventions.

First, Galileo's telescope. The telescope had a profound impact on the development of technology. Thanks to the telescope, humanity learned what the world was like. In other words, humanity didn't invent the telescope because it knew of the existence of the solar system; it learned about the world outside of Earth because of peering through the telescope. Similarly, I believe big data will help us discover a world we've never seen before. This is extremely important. Today, we are at a very early stage of understanding big data.

The second invention we can look at to understand big data is the microscope. The microscope can enable to you see minuscule things you've never seen before. An important event in the development of the microscope was the first time cells were ever observed. Things under the microscope are alive: even an immobile piece of wood is full of active cells. Through the microscope, people saw a world full of life that was invisible to the naked eye.

On the internet, there are many things that are often ignored or taken for granted. Before the internet, for example, no one knew that clicks had any special value or meaning. But today, everyone knows clicks can be given life and meaning through data and being online. They're like cells. Many cells together can constitute a life-form. The internet's most energetic and basic unit is the click, and it shouldn't be underestimated. Believe in a world you cannot see.

The third invention comparable to big data is radar. As World War II was ending, *Time* magazine planned to run a cover story in its August 20, 1945, issue about radar to explain the important role it had played for the Allies. But before the issue was released, the United States dropped two atomic bombs on Japan, and instead *Time* ran a cover showing a big *X* crossing out the Japanese flag. Most of the articles covered the atomic bomb, the people involved in its development and deployment, and the impact it would have on the world. The radar story was reduced to three pages. Nobel Prize winner Hans Bethe, a nuclear physicist who led the development of the atomic bomb, had conducted research on radar for two years while at MIT. He once said, "The atomic bomb ended World War II, but radar won it."

In the 1940s, the Royal Air Force (RAF) of Britain was very weak. In order to defeat the German Luftwaffe, they relied on radar to let them know where an attack was coming from, where the attacking planes were located, and approximately how many were in the assault. The RAF fighters would only be launched as the Germans approached British airspace. This meant that RAF planes could spend much more time in the air than their Luftwaffe enemies, which allowed the UK to repel the attacks by the superior German planes. Long-wave radar was one of the keys to victory in the Battle of Britain. Beginning in the summer of 1940, it successfully guided thousands of RAF aircraft in intercepting German fighters. Without radar, the RAF would have had to mobilize tens of thousands of planes, something they were incapable of doing at the time. The British very well could have lost if they had not had radar.

In September 1940, Henry Tizard led a team of scientists on the Canadian liner *Duchess of Richmond* that arrived quietly in Halifax, Canada. From there, they proceeded to

Washington, DC, by train. Finally, they met with Dr. Vannevar Bush of MIT to display the technology of the resonant cavity magnetron—a token of hope for British technology—and to share their radar technology with the Americans. The US military would set up the world's largest radar research lab at MIT, which later gave the United States a major advantage in radar technology against Japan in the Pacific War.

In Japan, radar wasn't considered particularly important, and for that reason Japan lagged far behind the American and British technology. Like the electrocardiogram of today that measures the electrical activity of the heart, contemporary radar would show a wave on the screen when there was an enemy aircraft, and when there was no enemy, it would register a flat line. But the device couldn't accurately locate the enemy craft. This type of radar wasn't considered an important piece of defensive equipment by the Japanese navy, which still relied on telescopes to spot attacking planes. But by the time you could see a bomber with a telescope, it was already too late. By contrast, the onboard radar systems of US aircraft carriers could track any Japanese aircraft within a hundred-mile radius.

In 1942, radar helped the United States defeat the Japanese at Midway Island, the Solomon Islands, and the Marianas and Savo Island. By 1944, the Japanese navy had won the right to use radar over the army, and they installed modern radar equipment on their warships. But at that point, they had lost their most important aircraft carriers, leaving them without vital air support. Near the end of the Pacific War, the Japanese navy dispatched the world's largest battle cruiser, the *Yamato*, to attack the American carrier fleet. It didn't take less steel to build this ship than an aircraft carrier would have required. Every shell it launched weighed as much as an automobile. For a battle cruiser, as long as its missiles had longer range and weighed more than yours, it could sink you. For the Japanese, the goal of sending out this ship was solemn and stirring. It was to engage aircraft carriers and destroy them. However, the *Yamato* didn't have time to find a single aircraft carrier because it was engaged by two hundred American airplanes. The gun turrets were so heavy that, when the ship capsized, they ripped themselves off.

Knowing where an airplane is ten minutes in advance is a huge advantage. This is the value of radar. Data can help you do that.

In the past, antivirus software would only react after a virus invaded the system. But if it could only react and not act preventively, your system clearly had vulnerabilities.

Microsoft discovered that if there was a computer virus spreading around the world, then someone somewhere six or twelve months previously had written a line of code. That code only told you where your system vulnerabilities were; it wasn't a virus itself. But this meant you could know where your system weaknesses were up to twelve months in advance and predict that a virus might possibly be created purposefully by someone.

After the internet had become infrastructure, people were able to do much of what they wanted online. All communications could take place over the internet. People must leave some trace of their activities, and viruses are no different.

In addition to the basic hardware and connective infrastructure, when I speak of the internet I'm including Bulletin Board Systems (BBS), blogs, messaging software, email, and Facebook, among other communication methods. As long as you process enough data, you can find the clues. Originally, you had to wait for the virus to appear to take action, but today if you analyze all the data on the Web, you can take control of the situation twelve months in advance. This is true competitiveness.

In the past, countries had to look at customs data to know trade statistics. But these figures are compiled long after the transactions themselves have occurred. Data from Alibaba's B2B service is different. You can predict what will happen based on the number of purchase inquiries placed on day one. If the number of buyers placing inquiries goes down, then exports will definitely drop. From this, you'll discover the value of data is far greater than providing a report.

If you know something before other people do, then nothing can get in your way. Even if you're ahead of them by just a few minutes, you'll still have a big advantage. Data can help you know things in advance.

As reported by the *New York Times*, while the 2020 presidential election campaign was still being fiercely contested in the United States, vendors in the Yiwu International Trade Market in Zhejiang province were trying to come up with their answer as to who would win in November. Four years earlier, when most political experts predicted Hillary Clinton would win instead of Donald Trump, many business owners in Yiwu predicted Trump would win even before most US voters cast their ballots, based on the fact that Yiwu had received four times more orders for Trump merchandise than for Clinton merchandise. The prediction was right in 2016, and people coined the term the "Yiwu index."

Could the Yiwu index correctly predict the US election result again in 2020?

The numbers showed that orders for campaign products for 2020 US election were significantly lower than for the previous one, but the market definitely sold more products related to Trump than Biden. The sales numbers failed as a predictor this time. Data doesn't lie, but data doesn't always work in a simple way.

Believing in Data Is a Kind of Self-Confidence

Whether or not you're willing to place the foundation of your success in someone else's hands is a question not just for cloud computing, but for systems, civic responsibility, society, and many other realms.

The greatest difficulty when it comes to promoting cloud computing and big data is due to the fact that they have already exceeded the limits of technology and common knowledge and have become closely related with human nature.

The greatest fear of banks is that someday people suddenly will cease to trust them. If everybody believed all at once that putting money in the bank wasn't safe, they would all withdraw their money. In many movies about the chaos in old Shanghai in the 1930s up to the late '40s, just as in movies about the Great Depression in the United States, there are scenes of bank runs. After the bank run, all of society undergoes upheaval. Not putting your money in your own country's banks and rather depositing it overseas is no longer just a wealth issue. It goes to matters of societal trust.

Bankruptcy of faith is a frightening thing. Cloud computing and big data are also dependent on trust. It doesn't matter how large the company is, it needs to have faith in the cloud platform, faith that the platform will not abuse it, or rip it off. But this alone is not enough. The most essential thing is earning the trust of people.

I once told my colleagues that to succeed with Aliyun, we need to earn the trust of entrepreneurs. Innovating and starting businesses on the cloud should mean that your company someday will also be able to list on the stock market and maybe even become bigger than Alibaba. The cloud can succeed only when there exists this level of trust.

Confidence in paper money

China was once the most creditworthy country in the world. A long time ago, precious metals were used as commodities of exchange, with value measured in weight. Words like "pound," "lira," and "ruble" all have their roots in units of weight as a measure value. In the history of money, paper money was a major milestone because its value was not in its weight but rather in credit. And with credit, trust is the true source of wealth.

The Chinese *jiaozi* is considered the oldest form of paper currency. It predates European paper money by seven hundred years. During the reign of Emperor Wu of the Han dynasty (140 BCE to 87 BCE), there were white deerskin coins, each worth 400,000 *qian*. The value wasn't directly linked to the actual value of the skin coin, so the deerskin coin was a precursor to paper money. By the Tang dynasty (around 800 CE) "flying cash" appeared, which consisted of notes that represented deposits and saved merchants the inconvenience of carrying around large amounts of heavy copper coins. In the Song dynasty (998 CE to 1003 CE), the first banknote considered legal tender, the *jiaozi*, appeared in western China's Sichuan province. *Jiaozi* is ancient Sichuanese slang meaning "voucher."

Here is a description of *jiaozi* from the Wikipedia of China:

> In the beginning, jiaozi were deposit vouchers. During the early years of the Northern Song Dynasty, there emerged businesses known as jiaozi shops in Chengdu, Sichuan Province, that made cash guarantees for merchants to make it more convenient for them to deal in large sums. The depositor would take the cash and give it to a jiaozi shop, which would then make a

receipt with special tissue paper and give it back to the depositor while charging a storage fee. This temporary deposit slip with an amount written on it was called a *jiaozi*.

As their usage became more and more widespread, many businesspeople joined together to formally issue the jiaozi as a currency and set up branch locations. Because there was trust in the jiaozi merchants, and you could get your money whenever you wanted, jiaozi became more reputable. In order to avoid the hassle of carting around copper coins, more businessmen began using jiaozi directly for large transactions. Later, the jiaozi shops discovered that only a portion of the deposits were accessed on a daily basis and this didn't affect the confidence in the jiaozi. Therefore, they began printing bills with the same amount on them as a first step toward circulating more jiaozi in the market. This step-by-step process made jiaozi gradually gain trust and acceptance and eventually it became paper money.

Jiaozi served as an official currency in Sichuan for nearly eighty years. The *Song History* reads: "The law concerning *jiaozi* originated from the flying cash used in Tang dynasty. Because the Sichuanese copper cash was heavy, transactions were inconvenient and a law was created applying to a currency. Each *jiaozi* was worth 1,000 coins, and they were traded in every three years. The 65th year represented the 22nd round of trading in. That is what is called *jiaozi*." From a commercial credit voucher to an official currency, it only took a few decades for *jiaozi* to fully develop, assuming the most basic elements of a modern currency. That people in ancient times could believe in *jiaozi*, could believe that a piece of paper with a stamp on it represented money that could be used to buy things, was a truly bold step. This was a sign of high confidence.

Subsequently, in the Tang and Song dynasties, people would come to believe in paper money when it became currency. Today we have gone backward. Because Chinese people don't believe credit cards actual represent an individual's credit, we often use them like debit cards. From this we can see that our ability to trust has decreased severely. Being willing to allow someone else to control the foundation for your success impacts our systems, citizen responsibility, society, and the environment. This is not just a cloud issue. In the long term, I'm optimistic, because I have enough confidence in data and I believe the course of history can be changed.

Confidence in cloud computing and data

Beginning in 2014, Midea Group, one of the largest consumer electronics companies in China, and Aliyun pledged to develop an internet-of-things platform based on the Aliyun cloud computing platform. Midea would use smart technology built on the cloud in its air conditioners, making the ACs internet terminals. The electronics firm planned to have half of its ACs operate on the cloud within three years. This had a profound impact on smart home appliances and the entire Chinese manufacturing sector. This case shows that it isn't only about having the ability to adapt but about a manufacturing firm's faith in an internet company.

In May 2014, Philips (China) signed a memorandum of understanding with Aliyun, having Aliyun cloud computing provide services and support on smart cities, smart health care, and the mobile internet. At the 2014 Frankfurt Lighting Exhibition, Philips displayed a smart lighting management system called City Touch, which could reduce energy use for lighting cities by up to 70 percent. Naturally, it drew a big response. Philips is a Fortune 500 company founded in 1891. This is a case of cooperation between a 120-year-old company and a firm that wants to last 102 years. By 2017, Philips (China) had moved its entire IT system to the Aliyun platform and closed its data center in China permanently.

This mutual confidence allows both companies to work together inside and out.

With the cloud, companies put their strategic resources, i.e., data, onto other people's computing platforms. This is like putting money in the bank. Thanks to cloud computing, data will accumulate quickly, to the point that it is practically too much to process. All of society must have confidence in this type of future.

Why must cloud computing come first before big data? The logic is simple: if there is no faith in cloud computing, big data cannot accumulate. In the past, people didn't have a deep understanding of the internet and just went online willy-nilly. As soon as they went online, they discovered it wasn't as safe as they had thought, and it became difficult to defend against online attacks. When it comes to cloud computing, people shift their concerns about the insecurity of the internet to the cloud. They're worried that letting their data be settled there will leave it vulnerable to some unforeseen harm.

But the development trend of society points toward further accumulation of data. If you take a walk to meet someone, you leave footprints, but if a strong wind blows, your footprints will be erased and you could never prove you walked that road. But if you take a bus to see someone, you'll have a ticket stub that won't be blown away in the wind.

The more footprints you leave, the more data you generate. As society progresses, there will certainly be things that are lost along the way. This process of progress cannot be reversed. But if you have the courage to embrace the future, your ability to be on the cutting edge will be strengthened. When Chinese people used *jiaozi*, China became a more advanced country than it was before paper money. But, since Chinese people today don't trust credit cards, compared to the countries that do use credit cards China is backward. When I was young, I had to pay the bill first to eat at a restaurant. Now you pay after you eat. Even though a few customers might eat and leave without paying, in general people have confidence in the current arrangement. This is credit, confidence, and trust.

I believe the internet can take our trust to the next level. If you believe standing in line will result in your getting what you want, then you'll form an orderly queue. If you discover that every time you stand in line, you don't get what you want, you won't wait politely anymore. Imagine you're standing in line at the bank and when it gets to your turn, the clerk says, "We're no longer seeing customers. It's time to close." Then the next time you're in the bank, you very likely might cut in line.

Around 1986, I was in Beijing, and at the time there were few foreigners studying at a handful of universities such as Peking University. The Beijing Zoo was a big bus transfer station that was packed with crowds of people. If you waited in line politely, you'd have trouble getting on the bus. I once saw foreign students, whom I assumed would be very orderly and stand in line, crazily push their way to the head of the line even more aggressively than the locals. In a situation of resource scarcity, order will be destroyed, and without order, then will be no trust.

If you do something once that requires your trust, and the result is that you get what you want, then you'll be more trusting in the future. If you didn't get what you wanted, then you'll trust less.

The online world in principle can satisfy the requirements that need to be satisfied. Because online resources are shared, if ten people read your blog, that doesn't mean that the eleventh won't be able to read it. This is different from newspapers. You don't have to worry that the thing you want will suddenly disappear. Previously rare resources are no longer rare. As long as everyone stops believing that it is hard to get things, they will stand in line, and trust will greatly increase. On sites like Taobao and Amazon, merchants want to trust you, and you want to trust the merchant. This kind of trust is closely related to the level of civilization in a society. Trust is slowly built up.

People working in the internet industry often discuss openness and sharing. At first, I didn't understand, but later I gradually grasped these two concepts. "Open" doesn't mean "free." In principle, it means everyone is equal. On the internet, everyone has access to resources. They cannot be monopolized by high-ranking individuals. Everyone is equal. The value of sharing is making resources hard to be controlled by a few, and even harder to be monopolized.

Take education, for example. Educational resources are the least equal and could be said to be monopolized. When Nicholas Negroponte was trying to provide computers to mountain villages, he discovered that many villages had computers that weren't being used because there was no electricity. So he developed a laptop that could be charged with a hand crank. From this example, you can see that education can be very unequal. Some schools have tried so-called multimedia education, but in many cases it ends up with students sitting in a classroom behind a computer connected to a local area network, watching the teacher go over standard courseware. There is not much value in watching the same teacher go over the same material in just a slightly different form. Something of actual value is making education go online, unleashing the world's educational wealth, and enabling it to be shared by everyone. As soon as education is online, China's western mountain regions will suddenly have access to the best teachers at Peking University. And the same is true for other countries with wide regional disparities in resources. The openness and sharing nature of being online should not become empty verbiage.

Sharing and openness in being online require truly cooperating with other people rather than trying to steal their meal. Only this way can both sides develop trust and avoid turning into adversaries.

Even if cloud computing disappeared one day, the trust developed would remain its greatest societal heritage.

Chapter 4

The Evolution from Computers to Computing

Computing Is a New Public Utility

The difference between things is sometimes determined not by their physical form but rather by the form of the service they provide. The cloud liberates computing from the PC box. A new form of being online is computing online, something everyone has heard of by now as cloud computing.

Why is computing called cloud computing? Why would I rather call it general computing? What's the difference? As an enterprise, you can supply your own computing needs as Google does, or you can view computing as a public utility and let a third-party cloud meet your demands, as with electricity. If you're like a traditional company of the old days and you buy your own power generator—in this case, your own servers—then not only do I think this is the wrong way of dealing with your problems, it also means your demand cannot be met past a certain point unless you are as big as Google. If you don't want to use the electric grid, there are two options: either you build a big enough power plant or, if you are a small company, you can use a diesel generator.

I remember going to Shenzhen once in the mid-1980s; two scenes left a deep impression on me. The first was: all the herringbone antennae, which were best for receiving the TV signal from the Hong Kong side, stuck on rooftops everywhere. The second was in the old shopping district, where every store had a small diesel power generator outside, so the streets rumbled with the noise of generators all day long. Why? At the time, China's whole electricity infrastructure had been built according to a rigid plan, and compensating for additional refrigerators and televisions was impossible. Imagine if computing were like the electricity infrastructure back then, and every time you added a computer you had to completely rearrange the network. It would be like using a generator on your own. So cloud computing can be viewed as two separate things: general computing and public utility computing. General computing is like electricity that can be used for an air conditioner or a fan. But only when computing becomes a public utility will a new era dawn. I think there are still many people who wouldn't agree with this position. Yet I continue to hold this opinion and move in this direction, even though I know it's going to be difficult and take a long time to accomplish.

The 2020 World Expo was planned to be held in Dubai. I visited the city in 2015 to see what cloud computing could do for this exposition. It's an interesting city, surrounded entirely by desert. As for the people, as long as something would be first in the world, they're willing to do it. I was speaking with an Emirati government official, and we discussed how to make the Dubai expo a "first-in-the-history-of-the-world" expo. Typically, large companies like IBM, in other words traditional companies, will provide the IT support for events like this. So we agreed that Dubai could become the first city to have public utility computing provide the computing power. I don't know if they'll accomplish this, but if they're interested, I'll work with them.

Why do this? At the end of the nineteenth century, electricity wasn't yet a public utility, but at the Chicago World's Fair running electricity was supplied to power the lights. What was displayed at the fair is probably long forgotten, but that the Chicago World's Fair of 1893 was the first to use electric lighting is something still remembered, because it was epoch-making. Because of the COVID-19 pandemic, the 2020 World Expo in Dubai has been postponed to October 2021–March 2022. I'm still hoping it will make history for cloud computing.

On January 19, 2017, a breakthrough partnership was announced by the International Olympic Committee (IOC) and Alibaba Group. With this partnership, it was agreed that the Aliyun would provide the cloud computing infrastructure and cloud services to support the Olympic Agenda 2020 and help the Olympic Games operate more efficiently, effectively, and securely through 2028. This is the first cloud computing partnership in the history of IOC. Traditional IT used to dominate the Olympic Games, and that has started to change with cloud computing. As the world has been grappling with the COVID-19 pandemic for more than a year, Tokyo 2020 became the first Games to be postponed in the 125-year history of Olympics. The year 2020 was hard for the world but also hard for cloud computing.

A brief history

John McCarthy was best known for his work in AI. He was the first author on the seminal proposal coining the term "artificial intelligence" in 1955. McCarthy also first proposed a concept of new computing in 1961 that is considered as the origins of utility computing: "If computers of the kind I have advocated become the computers of the future, then computing may someday be organized as a public utility just as the telephone system is a public utility. . . . The computer utility could become the basis of a new and important industry." Today, both cloud computing and data are the foundation of artificial intelligence.

In 1969, Leonard Kleinrock, one of the chief scientists of the original Advanced Research Projects Agency Network (ARPANET) project, also envisioned that: "As of now, computer networks are still in their infancy, but as they grow up and become sophisticated, we will probably see the spread of 'computer utilities,' which, like present electric and telephone utilities, will service individual homes and offices across the country."

In 1983, Sun Microsystems came up with the idea that "the network is computers." This was a precursor of cloud computing in industry.

In October 1997, University of Texas PhD Ramnath Chellappa spoke at the annual conference of the Institute for Operations Research and the Management Sciences (INFORMS). His key statement was that computing had already begun to shift from large machines to the internet. He called this computing model "cloud computing," using the term in an academic setting for the first time and providing the first academic definition of the term, which was known only by a very few people in industry. Later, as professor at Emory University's Goizueta Business School, Chellappa would continue his research.

On August 9, 2006, Google CEO Eric Schmidt spoke at a search engine conference in San Jose, California and discussed how Google was using the cloud and cloud computing to provide internet services. Schmidt emphasized that this was a very different model from the one being used at the time by Oracle and its traditional client/server processing model.

On August 25, 2006, Amazon launched the Elastic Compute Cloud (EC2) beta version. EC2 is the most important part of the Amazon cloud computing platform, Amazon Web Services (AWS).

The following month, Jeff Bezos gave the keynote speech at MIT's Emerging Technologies Conference, where he described EC2, S3 (Simple Storage Services), and Mechanical Turk as the combined result of their "eleven-year large-scale World Wide Web computing" services. In the speech, Bezos didn't directly bring up cloud computing as Schmidt had, but he repeatedly said that they would use back-end infrastructure as a service.

In December 2007, *BusinessWeek* ran a cover story headlined, "Google and the Wisdom of Clouds" that discussed Google's new strategy to put an incredible amount of computing power into the hands of the average consumer. This incredible amount of computing power, the article said, was what Google called "the cloud." From then on, the term "cloud" and even MapReduce, a programming model, became a media buzzword. Google's cloud computing services began formal operations in 2008.

In September 2008, speaking at Oracle OpenWorld, Larry Ellison stated that cloud computing was simply the latest fashion. The *Wall Street Journal* published the following report:

> The interesting thing about cloud computing is that we've redefined cloud computing to include everything that we already do. I can't think of anything that isn't cloud computing with all of these announcements. The computer industry is the only industry that is more fashion-driven than women's fashion. Maybe I'm an idiot, but I have no idea what anyone is talking about. What is it? It's complete gibberish. It's insane. When is this idiocy going to stop?

In November 2010, Amazon moved its retailers to EC2 and AWS, with Netflix and Pinterest moving their Web services to AWS as well. From then on, the cloud and cloud computing began to enter the mainstream.

In March 2010, Robin Li (founder of Baidu), Pony Ma (founder of Tencent), and Jack Ma (founder of Alibaba) were on the same panel at the China IT summit. One of the topics was cloud computing. Pony viewed cloud computing as something too early to invest in and much like the science depicted in *Avatar*. Robin bluntly stated that cloud computing was like old wine in a new bottle and nothing new. Jack was the only one of the three who believed cloud computing was the future and needed to be worked on. Ten years later, this conversation is still widely circulated in China.

Electricity as a public service

To differentiate between cloud and non-cloud computing, you have to look first at its core nature: is it connected to be online? Does the computing process require use of the internet to be carried out? This distinction reveals the two most important characteristics of cloud computing: first, it will become a public utility like water and electricity; second, as its computing scale grows, it will require massive data centers. From the first day I began working on cloud computing at Aliyun, I told myself, "Cloud computing is a fundamental public utility, just like electricity."

Baogang Steel, located in Shanghai, was the first modern steel factory in China. Its generator and the state grid both produce electricity, but are they different in any way? The answer is: there is a very big difference between them. Baogang's generator provides only Baogang with electricity. If there is a problem, it affects only them. The state grid provides power to countless homes, so a minor problem could have a big impact on average people's daily lives. For this reason, only the state grid's electricity can be considered a public utility.

Wind power and the Three Gorges Dam both generate power, but what is the difference between them? Ten years ago, the major portion of wind power couldn't be input into the state grid for economic and technological reasons, so wind power couldn't become part of the public utility. The Three Gorges Hydroelectric Power Plant was feeding into the grid from day one, so by contrast it was a part of the national public utility. In internet terms, you could call electricity that can hook up with the state grid "online power." Today, with the development of the necessary technologies, most wind power has become integral to the grid: online power.

Computing, regardless of its scale and computing power, won't truly be cloud computing unless it is online and part of a public utility infrastructure.

Why must cloud computing be a public utility? Because cloud computing is already the most basic societal necessity of the internet era, just as electricity was for the industrial age. Cloud computing will gradually become part of the national infrastructure and support the whole country's economic growth.

Imagine you were in New York at the turn of the twentieth century and you wanted to build a textile plant. Even if you had enough capital, workers, machines, and raw materials, you still couldn't begin to operate. Because in the United States at that time, there was still no public electricity. If you wanted to open a textile factory, you needed first to get your own generator. Back then, a generator could power only one house or a street of streetlamps, and only direct current was available. People called these generators "household power stations."

Once you had gotten your power generator set up and were preparing to turn it on, you had to think about one other thing: if your factory went out of business and you discarded the generator, no one would come and pick it up. Often, it was more expensive to build a small power station than it was to build the factory itself. And while building an electric power plant was much simpler than building a steam engine to run the textile factory, the low efficiency and high risk of this situation are obvious.

Humans are born with genes for doing business. People gradually realized that building their own power plants was truly foolish. If you didn't use all the power you generated, what did you do with it? Sell it to other people? If you didn't generate enough power, could you rent it from others? If two companies combined their power generators, it would be more effective, with lower overhead and risk. Why not link up a dozen or so power stations? Following this logic, more and more power stations joined together, and the impact of power at scale became apparent.

In 1882, the "king of inventors," Thomas Edison, built a power plant that used six generators in New York City. Even though this plant had only 30 kilowatts of capacity and could provide only enough electricity to power the city's streetlamps, it still marked the first time in history that electricity was used as a public utility. England's invention of the steam engine had set the foundation for their later greatness. By contrast, England had invented electricity long before the United States, but it was years behind the United States in using it as a public utility, which is one of the reasons Great Britain fell behind.

The emergence of public grids helped inaugurate yet another revolution in electricity technology. As more places tasted the sweetness of electricity and demand grew, the grid expanded its reach. Providing power to remote areas requires increasing voltage and lowering the power lost in transmission before being fed through a negative booster and linked to a house. Direct current (DC) electricity is very complex, whereas alternating current (AC) is not. From a usability standpoint, it's obvious that AC is better.

Electricity grids have become a national standard, and AC is now the world's current. Electricity has become an integral part of the infrastructure of our lives, society, and economy.

Today, it's hard to imagine a situation where you would have a power generator at your home routinely powering your lights. When the hot summer comes, if you decide to install an air conditioner, you have to make sure you're generating enough power to handle it, so you buy a bigger generator and install it on top of the old one. A few days later, if you want to buy a bigger refrigerator, you have to check whether you have enough space for another generator to power that. It's not sustainable.

Unfortunately, this obviously unsustainable scenario is making the rounds again, only now it is with regard to cloud computing. People are more interested in buying more computer servers than simply availing themselves of cloud services. Believing you can solve your family's power problems by buying more and more generators isn't just wrong, it's patently ridiculous. But that is what's happening when it comes to the view that cloud computing can be built just by building data centers.

Computing as a utility
In the internet era, everyone's demand for computing power is growing rapidly, seemingly without limit. Just as only a national power grid, and not private generators, could solve the country's electricity problems, cloud computing is the only way to meet the high demand for computing power brought about by the internet. This is an era of computing power being put online.

As far as hundreds of thousands of users are concerned, the cloud, just like electricity, has become an integral part of their lives.

The most common activity online is search. When we type the keywords into the search box and hit Enter, approximately the amount of energy required to power an 8-watt light bulb for an hour is expended, and it takes several thousand servers to provide computing power. Of course, this complex and frightening process is accomplished through the internet at a data center you know nothing about. Many of the things you do every day, including going on a shopping site, sending a post via an email service, sending a message, refreshing a news website, and watching a video all require an amount of computing power far exceeding that of your device. In reality, there is no computer in the world that could search the vast ocean of the world's data without using the internet.

In August 2012, typhoons Saola and Haikui struck China. Everyone in the country was going online to read news about where the typhoon was headed next. This caused the system to crash several times. Because many of the typhoon warning systems had too much traffic, Fujian and Jiangsu's typhoon information services crashed. At the same time, the Zhejiang Water Resources Department was able to support increased volume of visitors from all over the country, leading CCTV, Weibo and newspapers to use their data. In the past, the department had never had more than 50,000 hits in a day, and that was basically all from government workers. But as the typhoons approached, that number shot up to over 3.5 million. No wonder other systems went down. How was this possible? The department's servers were on the cloud, allowing it to easily adjust to the massive jump in traffic. Today, only the cloud can guarantee your life isn't suddenly pulled back fifty years to the time of the punch-card computer. People don't realize it because the cloud has become so basic. When it is working for you, you don't notice it. But when the cloud isn't working, you can clearly feel it.

From June to August 2012, sports fans had their hands full as the EuroCup and the London Olympics both enthralled audiences all over the world. More than 100 million Chinese people watched global sporting events live on the China Internet Sports Channel during this period. Traditionally, you would buy several servers in advance, but when it came time for the most popular events and games, you wouldn't have enough. Then, after the major events were over, you would have too many servers, and it would be a waste. For the London Olympics, China Internet TV Sports Channel used cloud computing for the first time, with being online providing the computing power. Not only did it guarantee stable streaming, it also saved them a lot of money. I watched a live feed of the opening ceremony of the London Olympics from Zhejiang's Putuo Mountain (one of the four famous Buddhist mountains on the remote island), and, because China Internet TV was using Aliyun, I was paying attention to the video quality and speed, not to the ceremony itself. What an experience! Two years later, CCTV's sports channel also used Aliyun for the Brazil World Cup. By then, I was able to relax and watch the games.

Why didn't the steam engine develop into the common power source? I believe it was because of the very basic issue of mobility. The steam engine isn't a mobile power source. You can't imagine dragging a steam engine near your house and using it to power your blow dryer in order to dry your hair after a shower. In reality, what you need is power, not an engine. Electricity is more mobile in the sense that you can simply connect a power line to wherever you are and you have power. But electricity's mobility problem now is due to its reliance on wires. Wireless electricity transmission remains the dream of many inventors. In July 2007, MIT's Marin Soljačić led a team that displayed wireless charging for a sixty-watt light from a distance of two meters. They called their system WiTricity and took a step closer to wireless transmission of electricity. Actually, the first person to display wireless electricity transmission was the inventor of alternating current—Nikola Tesla—a hundred years ago. And to this day people are still working toward the same goal.

The difference between servers and cloud computing involves a similar issue. You need movable computing power, not clumsy computers. The true way to experience cloud computing is to move it online and scale up like electricity. Having the ability to scale up depends on whether your company truly believes in being online and believes that online computing would be of benefit to society.

Now, many companies are looking into the cloud for education, health care, e-commerce, gaming, and many other vertical business sectors. One sector may sound better than another. But this is still thinking in terms of vertical applications and not truly investing in the cloud computing platform. There is no such thing as the vertical cloud. There are only vertical problems that the cloud can solve. Using the cloud to solve vertical problems is the right way to approach this. What can computers do? If you bought a computer to do word processing, then is it not just an office computer? If you bought it to play games, then is it not just a gaming computer? Or is it a computer that can be used for both gaming and office work?

Just as there are no vertical computers, there is no such thing as a vertical cloud. There is just a vertical application of cloud computing. Anything we call the cloud today is the application of cloud computing For this reason, the education cloud and the health-care cloud should properly be called cloud education and cloud health care.

I once gave a talk organized by the Chinese Ministry of Industry and Information Technology, where I stated that "99.9999 percent of data centers in China are not cloud computing centers." Because the data centers were just physical locations that rented out servers to other people, they could not be called cloud computing. Only a general computing platform can be called the cloud computing. A bunch of servers can't be.

Now many companies are working on cloud computing, but these platforms are either directly related to their previous business or are vertical solution platforms. No matter what, these efforts will solve only the problems faced by individual companies, not the computing services. Some companies that fly the flag of cloud computing are actually not providing a cloud computing service or building computing infrastructure. They're not making power plants; they're simply making custom-made generators. Again, a vertical solution is *not* a cloud computing platform.

Those vertical companies are essentially maintaining an attitude of charity: "I know what should be done, so I'll give you a platform to do it." This outlook must be changed and renewed. True creativity comes from the individual enterprises themselves. What they need is a fair and open computing platform.

When the first PC appeared, no one would tell the manufacturers what they should do with it. It was just a computing box. In this environment, the PC industry was able to undergo incredible growth. Now cloud computing is in a similar position as the PC when it was first invented.

I believe that to really do cloud computing, you have to use it as a common and comprehensive platform. When Aliyun was founded, I believed that General Computing Co., similar to General Electric (GE), might be a better name instead of Cloud Computing Co. That belief reflected my understanding of and faith in cloud computing.

The earliest form of computer network was a local area network (LAN), not the internet of today. LANs were networks to exchange data and information, and the other computers on these networks wouldn't help you do anything, so all of the computing tasks had to be done on your own computer. One scene in the history of computing is when people entered a series of commands into the command line of one computer, and a group of symbols was sent out by another computer.

The growth of networks led to the concept of servers appearing. Even though communicating with a server is essentially an information exchange, many things on the back end have already been processed. By the time of the internet, the most important work is being done on the back end, and the results of the computing are provided to the user in some form of a service. The servers everyone is used to are being replaced by the cloud now. On the surface, it looks like you've connected to the cloud, but in reality the computing happens online and the process brings back the computed results online. There is a distinction here with regard to the electricity analogy. Electricity can travel through the grid to countless numbers of households, and it is consumed locally. Cloud computing doesn't send the computing power to another location; it just takes the query to the cloud, and sends the results back to you. The computing is done by the computing facilities at the cloud data centers.

You can imagine it this way: when you search for something online, it is actually sent to a cloud that will activate thousands of computers to find the appropriate result amid a sea of data, and finally the result is presented to you. This is different from searching files and data as done in the past. The process takes up a considerable amount of cloud computing power; it's no longer a simple matter of sending and receiving data. This is what's called online computing. Online computing is something I've often emphasize in my public speeches by saying, "The cloud is the same as electricity." This concept will take some time to become widely accepted.

Internet users are already accustomed to the cloud, while traditional companies are falling behind. Many Chinese entrepreneurs working in the internet industry will begin by renting servers and setting up an internet data center (IDC). IDCs are built on the internet and provide maintenance bases and related services for intensive storage, processing, and data-sending equipment. Simply put, IDCs are the electricity factories of the internet age. If every company set up its own IDC, it would be no different from every factory installing its own generator back in old New York. It would be backward. Even if some companies reach a certain size, it is the same as building a massive generator next to your factory and not linking it up to the grid. It would still be backward.

If you make a website selling children's clothing and you use the personal generator model, then when no one buys clothes, your server is still running, and you are wasting resources because your excess computing power can't be used by anyone else. Then one day, if you have a sale, your whole server won't be adequate to handle the heavy traffic, and it will be too late to buy a new server. All you can do is watch customers trying to pay but being unable to. The most important unique characteristic of the internet is explosive growth. A mature traditional enterprise, like the IT department in an oil company, can make plans year by year. But it would miss out on many opportunities. That is changing now.

Companies in the internet age need to have flexibility built into their architecture.

If you want to resolve the problems brought on by everyone having their own servers, the only option is to turn cloud computing into a public utility.

In 2013, the tech executives community ITValue held its annual chief information officer (CIO) conference in Hainan. At the conference, I ran into the CIO of Jinjiang Star Hotels, a chain that was founded in 1996 and now has nearly 1,000 hotels under various brands. They were planning to rapidly expand the number of hotels they had. The CIO told me they had already built a large-scale data center but had stopped building a second one and were instead considering using the cloud. This is a particularly Chinese scenario. When encountering new technology, Chinese entrepreneurs are brimming with both enthusiasm and timidity. They rely on intuition to determine what's good and what's bad, are willing to take risks, and understand when to reverse direction. When I heard they'd halted construction on the second data center, I thought we were at a bold turning point in the history of cloud computing. That moment made me feel cloud computing was impacting the entire IT industry, and I wonder how Jinjiang has progressed since then.

Evolution from computers to computing was a small step for industry, but a giant leap for humankind.

There Is No Public or Private Cloud, Only the Utility Cloud

With computing as a public utility, if you think only in terms of selling hardware and software, you're wrong.

When I had just started doing cloud computing, I often said there are only one and a half cloud computing companies that Aliyun regards as role models. One is Amazon; the other half is Google. From the perspective of computing scale, Google is bigger than Amazon. But Google's computing resources are mainly used by itself, whereas Amazon is using the cloud as a public service.

I want Aliyun to be another one with public service to scale. A lot of companies have worked on the cloud to some extent but not fully. Embracing the cloud requires additional courage.

In 2006, AWS was born, and now the streaming giant Netflix, photo sharing leader Instagram, and many other big names in tech are using its services. Even NASA is a client. These websites are primarily using AWS to handle content storage and accessing, content selling, and purchasing services. Pinterest also used AWS, and was able to handle a jump from 50,000 hits to 17 million in nine months. Pinterest now receives 48 million hits per month. The company stored over 400 terabytes of content on AWS within that amount of short time.

The private cloud goes sour

Whether as infrastructure, platform, or storage, the cloud can be seamlessly integrated. This is the goal of cloud computing industry leader Amazon. I believe the greatest contribution Amazon has made in this regard is to help us understand that computing can be a public utility. Selling you a server is not the same as selling you computing power—these are two totally different businesses. Selling you a server means only that you have the ability to use computing power, but many servers end up rusting away in warehouses.

Computing power is actually hidden behind every industry. It is the foundation of the online society.

In his book *Code: The Hidden Language of Computer Hardware and Software*, Charles Petzold tells an interesting, but unverified, story. Today, there exist everywhere things that are encrypted. The encoding employs various sorts of encryption methods, which are set to make sure you can't break in during a certain time frame with a particular amount of processing power. The encryption methods of American banks are like this. Encryption methods must be approved by the National Security Agency (NSA). The NSA is America's most heavily protected intelligence agency, and it boasts the biggest budget and the most employees. The standard the NSA uses to evaluate encryption strength is how many fast computers are required to break it. People believe that only the NSA has computers fast enough to engage in high-level codebreaking, and hackers couldn't possibly possess that level of computing power.

Regardless of what technology people are talking about, the core of its operation must be the computing power. By providing computing power services, Amazon is making a major contribution to the positive development of the whole industry and is thus deserving of respect.

What level of cloud computing services can be considered a quality public utility? To go back to the electricity analogy, the criteria should be sufficient supply, low cost, and safe. When you plug in to a power socket, you'll immediately have power, and if you want to plug in ten electronic devices, there will still be enough power. If electricity is very expensive, using it for a minute could empty your wallet. But electricity shouldn't put an economic burden on individuals, so it must be low cost. And it wouldn't be acceptable if, when you're using it, you're not protected from being electrocuted. It has to be safe. Safety is the hidden foundation of all public utilities.

These three characteristics seem simple, but they are difficult to achieve.

Starting in 2006, industries started broadly to discuss and use cloud computing. More than a decade has passed. Many people have grown accustomed to only talking about it rather than doing it or even contemplating what it truly means.

Nowadays, many companies will talk about how wonderful cloud computing is and how terrible it is if you aren't using it. But if you speak with them directly, the discussion will be a little strange: "You want to use cloud computing, right? Great. But the public cloud isn't safe. If you buy my servers and database and storage services, I'll build a private cloud just for you." People often ask me what I feel about the public cloud vs. the private cloud. My response is, "I've never actually heard of the private cloud." That's a tactful response. Objectively speaking, the private cloud doesn't exist.

Around twenty years ago, if you wanted to encourage companies to grow their internet business, who would you have had in mind? It certainly wouldn't have been the internet companies we're all familiar with today. Most likely it would have been AT&T or some other carriers. AT&T might have responded that it was in the internet economy already because its network extended all over the country. Everyone knows now that the internet is different from a telecom network. The telecom network could continue to provide small-scale services for particular organizations, while the internet economy grew naturally. The companies that promote the so-called private cloud claim to be in the cloud computing industry, but actually they're just like AT&T claiming to be in the internet business. Cloud computing is a public utility. If you're simply selling hardware and software, then there won't be a computing revolution; at most we'll just see an extension of traditional technologies. The so-called private cloud is akin to a transportation company building its own highways. That sounds ridiculous on its face, because it is.

The supporters of the private cloud use a very simple and powerful argument: the public cloud isn't safe. The private cloud has exploited public concerns about security to sell people hardware and software. Everyone on the internet faces the same security issues. Unless you create a local area network and don't connect up with the internet, you won't be guaranteed security. But a LAN won't meet the needs of today. For example, when people talk about banks and the internet, everyone's first reaction is that banks are safer than the internet; the internet's infrastructure is by its very nature unsafe. But in fact, it's just the opposite.

The safety of the public service
While I often use electricity as an example to explain my thoughts about the nature of cloud computing, something the vast majority of people don't know is that when electricity was first becoming a public utility, it elicited apprehension and concerns similar to those elicited by cloud computing today.

Nikola Tesla was an inventive genius, a legendary figure. When he emigrated from Europe to the United States, he only had four cents in his pocket. He was almost penniless when he passed away in 1943. But today everyone in the world enjoys the benefit of his inventions.

In 1884, because he admired Edison, Tesla, a Serbian engineer, went to New York to work for the Edison Illuminating Company for eighteen dollars a week. New York was the Silicon Valley of its day, a haven for entrepreneurs. Edison was a strong proponent of direct current, and he had built the first public utility electric plant in New York providing direct current power. But there was a big problem with direct current. It had limited transmission range, at most about three miles. One day, Edison gave Tesla the problem of improving direct current plants and generators and offered him $50,000 (the equivalent of $1 million today) if he could solve it. After a few months, Tesla successfully made the improvement, and when he went to Edison to ask for the $50,000, Edison said, "You don't understand our American humor." Tesla quit on the spot after refusing Edison's offer to increase his salary by ten dollars a week.

After leaving Edison's company, Tesla set up his own laboratory and began researching alternating current from generation to transmission, which proved very fruitful. At the time, while direct current electricity couldn't be transmitted over long distances, alternating current could, by using transformers to change the voltage for more efficient transmission. Additionally, the cost of direct current was very high and the efficiency far lower than that of alternating current.

In 1887, Tesla launched the Tesla Electric Light & Manufacturing Company and completed his greatest invention—frictionless AC power generation—and earned a patent for it. At the time, AC transmission costs were one-thousandth of those for DC. In 1888, Tesla met George Westinghouse, the founder of the eponymous company and the inventor of air brakes for trains. Westinghouse paid Tesla two dollars and fifty cents per horsepower for the right to use his AC patent. By 1890, the thirty-four-year-old Tesla was a millionaire.

Edison obviously didn't want to see DC come under threat, so he began putting pressure on Tesla. It's hard to imagine now, but there was a fierce battle over which current would become predominant, and it wasn't among consumers. It was between the two inventors.

Edison's strategy was to create and spread stories about the potentially fatal danger of using AC. In West Orange, New Jersey, he brought several mammals in front of a live audience and activated a 1,000-volt charge in order to electrocute them to death. This was how Edison chose to persuade the public that AC was unsafe. The media covered the event live, and the word "electrocution" was invented to describe what had happened. In what was termed the "current war" between AC and DC, there were two major events: one was the decision to use the electric chair to execute convicted criminals; the second was the electrocution of the elephant Topsy with AC.

In June 1888, the New York legislature made electrocution a legal method for carrying out capital punishment but didn't choose whether to employ alternating current or direct current. To design the electric chair, Edison hired inventor Harold Brown along with his assistant Fred Peterson, who happened to be on the government commission to determine which current to use. Since Peterson was an employee of Edison, the decision to go with AC isn't surprising. In January 1889, when the first lethal electrocution was to take place, Westinghouse refused to provide AC power to the judicial authorities, and in the end it was Edison and Brown who supplied the AC for the electrocution. Ironically, people nonetheless referred to electrocution as "being Westinghoused."

In the first half of the twentieth century, New York's Coney Island was famous for its amusement park. Topsy was an Asian elephant in the circus there, and because she had been habitually abused, she killed one of her trainers. Now considered a threat to public safety, she was going to be hanged, but animal rights groups sought an alternative. Edison proposed that AC be used instead to put her to death, and around 1,500 people witnessed the execution. Edison's company filmed the event, and he had the motion picture played across the United States to show how unsafe Tesla's AC was.

In his campaign against Tesla and AC, Edison played on people's fears of the unknown, made those fears seem concrete, and magnified them. This was a tragedy that couldn't be avoided in the course of human development.

Westinghouse would later provide Tesla with more capital to keep researching. Finally, in January 1893, at the opening ceremony of the Chicago World's Fair, Tesla's AC electrified 90,000 light bulbs at once, shocking the whole crowd because DC could never provide that much power at such a low cost. Afterward, Tesla was chosen as the electricity designer of the Niagara Falls hydropower plant, and from that point forward, AC replaced DC as the main source of power.

Tesla remains the idol of many entrepreneurs. It is said that Larry Page, one of the founders of Google, read a biography of Tesla when he was twelve and decided to be an engineer. And when PayPal billionaire Elon Musk was setting up his electric car company, he decided not only to adopt the name of the Serbian inventor but also to use Tesla's invention, AC, to power his cars as well.

History's repetitions make me both pessimistic and optimistic about cloud computing. I'm pessimistic because I worry that a meaningless conflict, like the one between AC and DC, may arise. But I'm optimistic because it is possible to determine how true cloud computing and online computing power should be used.

In reality, the battle between AC and DC was about whether or not the electricity infrastructure needed to be made uniform. The telecommunications industry once operated on the concept of a private network, which was a large corporation making its own closed network while designing special security measures for its protection. Private networks had the ability to link to the internet, and they relied on the same infrastructure but their security systems were each different. Now the so-called private cloud and the public cloud are two different infrastructure systems. In order to guarantee a uniform computing infrastructure, just as with electricity, the public cloud's infrastructure can be made as secure as private clouds. But the two must be part of the same infrastructure.

Putting your confidence in cloud computing is the only true investment. Scaling up is a necessary consequence of cloud computing. Scaling up computing isn't just something that *can* be done; it's something that *must* be done. Scale is the only thing that can resolve cost problems. The effect of 90 percent of a country's electricity being on the grid versus 1 percent is totally different, at the very least in terms of cost, which can impact the penetration of a technology.

Apsara: Driving Computing to Be a Public Service

The combination of two things can result in a chemical reaction out of which something new is born. Cloud computing is one part of a chemical reaction.

Cloud computing is a new type of infrastructure for the twenty-first century. In terms of its overall importance, it's more than just one more business. Building such an infrastructure requires both belief in it and technology innovation.

Profoundly believing in what you're creating is critical for technology innovation. Professing love for something you actually fear can be a disaster. Some vendors say in public that cloud computing is a service but then privately sell hardware and software to clients while telling them to create a private cloud. This approach will never bring you true technology innovation.

The dream of Apsara
Apsara is the software specifically designed with the purpose of "computing as a public utility." (*Apsaras* are sculpted figures,

which the Mogao Caves in Dunhuang, China, are famous for, and they're usually referred to as Flying Apsaras.) It has a newly designed architecture to support what I call data-centric cloud computing. It is the software foundation of Aliyun.

IT Manager World magazine reporter Lei Wu interviewed me in 2013 regarding Apsara and cloud computing.

"Cloud computing changes the service model. This is a familiar statement," said Jian Wang. "But people often forget the technological difficulty faced by cloud computing, which is not minor. When we decided to scale up, we could no longer rely on other people's technology." Wang struck on the idea of innovating indigenously a large-scale distributed computing system. He called this platform Apsara, the design of which started in the second half of 2008. On the first day of work after the Spring Festival in 2009, in an old office without an air conditioner located in a northern Beijing suburb, the Aliyun team wrote its first line of code.

In 2009, I went to Silicon Valley, and in a speech at Stanford University I first shared the basic idea for Apsara. Many people felt that making Apsara in China was unreasonable. I agree. No one in the world would believe that developing an industry-standard, large-scale distributed computing system like Apsara in China was reasonable, because it was so difficult, especially at that time. But the interesting thing is that many people who initially had doubts later joined the Aliyun team and worked on Apsara. To work in cloud computing, you need to have at least a little bit of an adventurous spirit; otherwise you won't do well.

On a certain level, Apsara is an operating system for data centers. An operating system's most important feature is its ability to manage resources. Apsara is based on data centers with tens of thousands of general servers, each featuring multiple central processing units (CPUs), memory, and hard drives, and all linked together with Ethernet cables. The Apsara system, simply put, makes it seem as though all of its resources are in one huge computer, and it provides computing services through the internet.

Apsara is the name of a cloud spirit living by water in the legend of Mogao. She brings people fortune and good luck. As with the name *Apsara*, many modules within the Apsara distributed computing system have been given the names of ancient spirits of China. For example, the distributed files system is named after the founder of all, the one who opened the heavens and split the Earth, Pangu. The module that runs the task manager and research manager is the god of prognostication and the mythical author of the *I Ching*, Fu Xi. The diagnostic system is called Hua Tuo, after a legendary healer. The module responsible for linking up the network is the walking giant Kua Fu. The surveillance system is called Shen Nong, famous for his ability to taste whether herbs were poisonous or not. Cluster deployment is Da Yu, celebrated for building a river network for locals. All of these mythical names reflect the idealism behind these modules' development.

With the assistance of the gods and legends, Apsara is responsible for managing the physical resources of the data centers' Linux clusters, controlling distributed programming operations, and fixing hidden lower-level flaws and data redundancies. It effectively provides flexible computing and load-balancing services. Further, with thousands of nodes, the platform's various modules continue to be improved and optimized for scale, usability, and maintenance.

The Apsara system is an organic entity at a very large scale. The biggest challenge of the large scale is that there can be malfunctions in this "supercomputer" every day. The hard drive could fail, the fans could break, the memory processors could cease to function. You might be able to buy a laptop computer and use it for five or more years without any problems, but the thousands of Apsara clusters contend with malfunctions constantly.

After almost three years of tackling key technological problems and after countless updates and redoes, Apsara was finally completed and launched in fall 2010 at the first Aliyun developer conference. The computing and service framework Apsara supports has a more flexible programming model than MapReduce (a programming model used for concurrent operation of large-scale data clusters). Further, this single platform can operate email, search, maps, flexible computing, and data processing services, among many other functions.

Apsara's ground-floor structure is very different from Amazon's. Every service Amazon provides corresponds with one or many specific data clusters. For example, cluster A handles S3, Simple Storage Service, while cluster B does ECS, Elastic Container Service. Their functions and system design are vertical. Apsara's services all correspond to the same system core and the same distributed file system. In other words, whether it is storage, flexible computing, email, or search, they all operate on the same ground floor. Aside from Aliyun, Google is the only other company in the world that has been able to accomplish this.

Apsara's creation makes me very proud of my team. We are the first in China to have a general computing platform at this scale, including single data clusters with more than 5,000 servers, a general computing platform at this scale. Further, the Apsara team wrote all of the code ourselves, even though the barrier to entry for this industry is very high.

The cloud is services on Apsara

I discussed this service model in detail in the same interview with Lei Wu in 2013:

> "Can you stay open to provide services forever?" a client once asked Jian Wang. He responded, "I promise you will be the last customer I serve." Although this was in jest, the question made Wang realize that "cloud computing is a business based on trust." More importantly, the Aliyun team from then on began to understand the true meaning of service.
>
> Aliyun's first clients within Alibaba Group were Alibaba Finance, and then search, and email, and all began using Apsara in 2010. This symbolized that the Apsara system had formally launched. On July 28, 2011, Aliyun's official site (www.aliyun.com) went online, and formally began providing services to the public. The first product sold was elastic computing.
>
> Today, Aliyun's clients number more than 100,000, and Apsara has handled more than 100 million client hits. The client list includes Lenovo, Schneider Electric, Skyworth Television, China Internet TV, DeNA, and other major Chinese and foreign corporations; many mobile internet companies including Xiami music streaming services, Changba, Weekend Illustrated, and ifanr, and Alibaba's own Taobao and Tmall, Alibaba Finance, and the group's data platform, Binghuoniao.
>
> The road, however, has not always been smooth. Jian Wang and his team encountered serious doubt, skepticism, and criticism both inside and outside of

the group. But for Wang, these criticisms were not "a one-sided interaction. It wasn't like some people were telling us that we didn't understand what we were doing and others said our technology was bad." Aliyun maintained good communication with its clients the whole time.

Wang frequently says, "Clients taught Aliyun how to do service." I have heard that because the operating team is concerned about Wang's spot checks, they often go over client feedback and issues and ask the staff how problems were resolved. This has put pressure on their subordinates. Additionally, Wang also looks at customer feedback on Sina Weibo every day. Once, a client using Aliyun to render a 3D movie posted on Weibo that everyone should support Aliyun. This deeply moved Jian Wang, who responded by saying, "It is the client's support that makes Aliyun grow and improve."

On some level, Aliyun is an explorer. The tuition for every step of its journey it has paid itself. Wang gave himself an incredible goal. In order to reach this goal, he gave up all rest and works from 9 a.m. to late in the evening, sometimes into the early morning. His driving force is his belief that Aliyun is doing the right thing. "Sometimes persevering is difficult, and sometimes giving up is just as hard. It is always possible to be stuck in this kind of dilemma. But when it comes to Apsara, I believe persevering was much harder." said Jian Wang.

Clients are gods

After laying out how meaningful cloud computing is, there is one more thing I'd like to say: cloud computing is not important in itself. As soon as cloud computing becomes a public utility, the really important things will have nothing to do with cloud computing. What matters is its application. You can imagine it this way: In the summer you come home and turn on your air conditioner. You open your refrigerator and take out a cold beverage. Then you turn on your TV and watch a sci-fi movie. You take out your cell phone and look at your friends' posts on WeChat. You appreciate the amenities of modern technology and how they changed your life, but you would never think of the power grid. This attitude is normal and widespread.

In *The Big Switch*, Nicholas Carr writes:

> You type in a keyword, and Google's network of hundreds of thousands of interlinked computers sorts through a database of billions of web pages, draws out the few thousand that best match up with your keyword, arranges them in order of relevance, and shoots the results back through the Internet to your screen—usually in less than a second. That amazing computing feat, which Google repeats hundreds of millions of times a day, doesn't happen inside your PC. It couldn't happen inside your PC. Rather, it happens miles away, maybe on the other side of the country, maybe even on the other side of the globe. Where's the computer chip that processed your last Google search? You don't know, and you don't care—any more than you know or care which generating station produced the kilowatts that light the lamp on your desk.

In fact, electricity is not some amazing thing. It only became important because of washing machines, refrigerators, and TVs and other electronics. The electric grid and computing going online are both foundational—who knows what will grow on

these foundation? Perhaps when you're reading this, there is a
light on in your room. Even if it's nighttime, the room is still
very bright. This means the cost of electricity isn't something
you think about; otherwise you wouldn't be reading at night.
Cloud computing lowers the per unit cost of computing so
much that you don't even think about it so you can focus very
much on your innovation

Changba is an online singing app. The company is very
small; there are only forty or fifty employees, but it has more
than 40 million users. Changba makes karaoke go online and
liberates it from the constraints of time and space. This is something that has had a big impact. They rely on the cloud; otherwise
they couldn't support so many users anywhere across China.

There are many examples like Changba. They are all
branches sprouting from the tree of cloud computing and are
things no one could have imagined before.

Cloud computing is the most fundamental service. It
builds a platform that makes life easier for innovation and
start-ups. Cloud computing can make traditional industries
turn into internet industries, let a small company compete
directly with a large one, and transform internet companies
into mobile internet companies. Cloud computing allows us
to see that companies aren't divided between large and small,
but rather any size company can create unique innovations.

I once said the biggest regret for Chinese internet companies and Chinese research organizations is that we don't have
the ability to gather and search all the world's information on
the Web. I believe that once cloud computing reaches a certain
level of growth, if you want to start to build a search engine,
all you'll need to do is find the best people and put in a sufficient amount of effort, and then you can use the cloud infrastructure to support you. You won't have to be like Google
and build your own computing infrastructure.

Cloud computing can let you reach a scale you never could have before and do things you couldn't have done before.

Innovation started from an American engineer

Long ago, when you watched movies, you could do it only at night because you had to wait until the sun went down to see the picture clearly. And you also had to wait until winter, because in the summer, a room full of people would be too hot. Later, air-conditioning was invented in New York, and it changed the way movies could be enjoyed.

In 1925, Willis Carrier convinced Paramount Pictures to install an air-conditioning system in its theater in Times Square, and he promised it would keep the audience cool. When the theater opened, it became a big success, as hundreds of people flocked there, many attracted simply by the air-conditioning itself. The cool sensation completely won over the United States. Paramount president Adolph Zukor came from California to experience it. People had suddenly discovered that during the summer, when there was nowhere else to go to beat the heat, they could go to the movie theater and enjoy both the air-conditioning and the films. Before then, theaters never opened during the summer, but overnight, they became the biggest summer entertainment venue.

This was an important turning point in the development of movie theaters. If you don't understand this detail, it's hard to imagine that air-conditioning, which has nothing to do with movies, played such a crucial role in the golden age of cinema. In five years, Carrier installed air-conditioning in more than three hundred theaters around the United States.

In China's second- and third-tier cities, there used to be many viewing rooms that could seat only a few people and would play Hong Kong and Taiwanese movies. Young people loved going to those places. If you were paying attention, you would have noticed that on the doors to the viewing rooms, there always hung a sign: COLD AIR IS ON. Like movie theaters, these places did their best business in the summers because of the air-conditioning.

This story about air-conditioning shows how two seemingly unrelated things combine, spark a chemical reaction, and then create something new. Cloud computing is the same. It is one part of a chemical reaction that will result in more changes and innovation.

Believing in cloud computing requires having faith that your customers will use your platform to make even better things. Cloud computing is a foundational guarantee. An electricity company should not think about making refrigerators, just as a cloud computing company should not think about making apps itself.

I believe if cloud computing service providers make more money than the companies that use their services, then something has gone wrong. A five-star hair stylist can earn hundreds of RMB just for blow-drying hair. But how much does the state grid earn? Obviously, just the small electric fees. If the state grid thought otherwise and decided that it should earn a commission, then the hair-care industry would be in a complete mess. Once people make the cloud equivalents of refrigerators, televisions, and washing machines, then cloud computing will be truly valuable. That's why I say the cloud is critical but not important. The important things are the air conditioners and the blow dryers.

I firmly believe that in the cloud computing era, small companies will be able to do what only large companies could have done in the past.

First, computing will become a public utility. In the past, electricity was a rare resource that was difficult to access. Now it has become a public utility. In the past, the capacity of your online services was determined by your computing power. Now the cloud has expanded access to computing power. Today you can use a large company's computing power and a start-up's innovativeness to provide a massive service.

Second, in the past, attracting users was a very high-cost activity, but after the spread of the internet it became so cheap that your product could be adopted by everyone practically overnight. This would have been impossible in the offline world.

Cell phone apps grew so quickly because they could reach their users in an online way, so any innovation could quickly generate value.

Cloud computing can turn a traditional company into an internet company. This reduces IT costs, and, after it is on the cloud, the company can rethink its core businesses. I greatly respect the fact that so many enterprises have faith in us and put their businesses on our platform. Today companies that are bold enough to embrace cloud computing deserve admiration. Cloud computing provides companies with an unimaginable ability to innovate and space to develop. This is a new revolution.

Innovation and being online have a mutually reinforcing relationship. But they must be built on a foundation of technological development. I am fortunate to have personally experienced that period of burning passion.

De-IOE: Computing Is Replacing Computers

In August 2013, *Business Value* magazine ran several articles based on interviews with me by Yuting Zhang concerning the practice and development of cloud computing. I mainly talked about how cloud computing was going to revolutionize traditional IT infrastructure. "De-IOE" had become a hot topic in the industry that year, and this phenomenon reflected the irreversible trend of computing to phase out replacing computers. "De-IOE" refers to removing the reliance on IBM minicomputers, Oracle databases, and EMC storage, which are iconic symbols of traditional IT.

The origin of de-IOE

After joining Alibaba, Jian Wang, a technologist by nature with an academic flair, raised the idea of de-IOEing the company and starting a cloud computing push that could be written into the company's IT DNA.

These missions are helping move Alibaba from commercial software to open-source, with the ultimate goal of creating indigenous technology and cloud computing services. This also laid the foundation for Alibaba Cloud Computing Ltd. Co.(Aliyun), founded in 2009. On July 28, 2011, Aliyun's Apsara cloud computing platform began providing commercial cloud services as a public cloud.

Wang's office is full of books, with *Big Breasts and Wide Hips,* a novel by Nobel Prize winner Mo Yan, *Steve Jobs* by Walter Isaacson, and *Principles and Models of Distributed Systems* stacked on the third shelf. It would be appropriate to use these three books to summarize Wang's personality and career. With his background in engineering and PhD in psychology, he leads his team with a magical realism that lets dreams become a reality. All of the products and modules within Aliyun are named after mythical and legendary figures, from their core technology engine Apsara on down. . . .

In November 2008, Wang joined Alibaba Group and constantly moved between the positions of CTO and president of Alibaba Cloud. One of the walls in his office is almost entirely covered by a design drawing for YunOS, Alibaba Cloud's cloud-based mobile operating system. It has already become normal to hold meetings with his team in the Zhong Kui Problem Solving Room until late in the evening. He has both the sensitivity and the stubbornness of a product manager.

Switching between being CTO and the head of Aliyun was a huge challenge. He needed to find and develop the key core technologies for Alibaba Cloud. From indigenously innovating cloud computing technology to developing the YunOS operating system, every decision was of the utmost importance, and any setback could be devastating. But he always rebounded even after being widely criticized in and out of the company, and he was able to keep moving forward by solidifying Aliyun's tech base and accumulating more technology experience.

Wang said that most of his work was as CTO but he always thought, "I am extremely lucky, because without Aliyun, Alibaba's technology would lack a very important space to grow in. If Apsara had not entered the market, and we didn't have a chance to connect with consumers, we would not have been able to develop such truly good technology. Also, without Alibaba's other businesses, many of our cloud computing businesses, like our content delivery network, would not have been possible."

In Wang's opinion, cloud computing is a revolution that uses services to overturn the traditional IT and software era's attitude. Jian Wang believes that the de-IOE process completely transformed Alibaba's IT structure, allowed Alibaba to embrace the cloud, and produced a new foundation for computing services. De-IOE is in essence redistribution that makes it possible to buy commercial computing power anywhere. It is a precondition of making cloud computing work.

The de-IOE process was followed up by adopting open-source technology, moves that are paving the way for a complete migration to the cloud for all of Alibaba Group's IT. These successes have greatly added to Jian Wang's stature.

Pay to use computing, instead of buying computers
IOE are a product of the software era, or what could be termed "the computer buying era." Cloud computing, on the other hand, is the era of "buying computing."

Others believe the reasoning behind de-IOE is different. Wang is worried that the conversation around de-IOE can be taken to two extremes. One is de-IOE simply cuts costs. The other is the matter of being able to see if the company can go without foreign technology. A drop in costs is the first and most visible impact of de-IOE, but the root reason is because in the internet era, not just internet companies but the vast majority of enterprises have computing demands that cannot be met by IOE technologies alone. IOE restricts the growth and development of companies. Relying exclusively on hardware equipment owned by a handful of companies is rather dangerous, while being able to purchase commercial computing resources anytime anywhere is the safest route for Alibaba and many other companies. Jian Wang said all of the people talking about open-source technologies only mention how open-source solves the problem of the high cost of using software. But they ignore the costs of upgrading and maintaining open-source software.

In theory, as long as you have enough computing power, you can get rid of IOE. But in reality, the de-IOE process has considerable technological hurdles and risk. For the vast majority of companies, de-IOE is not simply changing software and hardware, replacing the old with the new. Rather, it is about using new methodologies to replace old ones; using cloud computing to completely transform your IT infrastructure.

From the first day he raised the idea of de-IOE, Wang has not stopped pursuing this goal to the point that these activities have ushered in a whole new attitude at Alibaba. It has also inspired other large enterprises on transforming their IT infrastructure, which forced IBM, Oracle, and other global companies to face new pressure brought on by the transition. Aliyun, on the other hand, has become Alibaba's main point of contact with many small and medium-sized enterprises.

Technological evolution: commercial, open-source, and home-grown

Alibaba originally relied on commercial software, using over twenty Oracle RAC data clusters (the largest in Asia at the time). But now it has transitioned to developing its own software with open-source MySQL, and its home-developed database OceanBase is used in several different businesses. It has moved from Oracle's data processing, to Hadoop clusters, finally to its own Open Data Processing Service (ODPS) based on Apsara. This path shows that commercial software, open-source software, and indigenous technology are always complementary. As far as different enterprises are concerned, it is just a matter of percentages. For a large company, indigenous innovation is very important. Alibaba has many examples, one of which being Apsara, the core platform that opens cloud computing up to users, thereby making cloud computing a solution that suits many more companies than before. "I believe that in the future the combination of cloud computing, indigenous technology, and open-source technology will become more and more important to many companies," says Jian Wang.

Alibaba's technology has developed from commercial software, to open-source, and finally to its own, in this case, cloud computing.

In many situations, people are accustomed to thinking open is the same as open-source, but open-source is not the only way to be open. In the internet era, cloud computing is a new way of being open. For many businesses and apps, open-source is not as good as cloud computing. Open-source software is still software, meaning it is a product of the software era and requires relatively high investment for maintenance and upgrading. Today's cloud computing benefits from the development of open-source software but at the same time challenges the applications of open-source software. For example, when cloud computing can provide relational database service, you will reconsider whether you need to buy commercial or open-source database software. Aliyun, CODE. CSDN (development services platform), and Open Source China are cooperating and hope the open-source community and its technical ability can take advantage of cloud computing and jointly serve users.

In the internet era, running software in data centers creates complexity when problems are encountered. There is a very big difference between using software (including open-source) and running data center software, and most companies don't yet have this capacity. The division of labor in society doesn't demand that every company become capable of this type of service. Cloud computing has given companies an opportunity, allowing them to enjoy great service without buying software and at the same time letting them have computing power. That is why cloud computing platforms, indigenous innovation, and open-source software will form the foundation of many companies, while traditional companies that rely on collecting service and licensing fees will continue to lose market share.

How can we resolve users' concerns over the cloud? Safety concerns are essentially an issue of trust. The only thing that can resolve them is transparency, and for that we need a transparent system. Today, many of Aliyun's clients are migrating from Amazon, while some clients unhappy with Alibaba's service have moved elsewhere. Cloud computing has staying power, but now there is an openly competitive environment and no one has the ability to restrict anyone else's activities. Aliyun also provides migration tools. Wang believes this can be discussed in concert with trust issues, and there needs to be a public standard.

Cloud computing is a revolution that provides value to innovative clients that far outstrips its costs.

The practice of De-IOE

2012's November 11 Singles' Day sale produced 19.1 billion RMB worth of transactions on Alibaba. Even though that hardly compares with the 91.2 billion RMB on 2015's Singles' Day, it had special significance for the cloud. the Aliyun Apsara platform processed 20 percent of Taobao and Tmall's sales and handled the massive increase in hits with zero interruptions and no lost orders. On the cloud, there are Tmall sellers who have received nearly 600,000 orders, marking the highest transaction volume on a single day of any other Chinese B2C platform. Before Singles' Day, many vendors didn't have faith in Aliyun, and they still used their own IT infrastructure. When the transaction volume exploded and their own servers couldn't handle the traffic, Aliyun helped them move over to our platform in under ten minutes, which guaranteed there would be no delays in their transactions. Some companies moved over to the cloud as late as midnight.

In the past, when a transaction connected from Taobao to the seller's own back-end system, if their IT infrastructure was weak, the transaction could fail for a variety of reasons, including connectivity issues. There was an expression that came out of Singles' Day, "20 percent of the transactions were completed over the cloud." It actually means that sellers who accounted for 20 percent of the transactions had their back-end systems deployed on Aliyun. When a buyer would purchase something, it would take one click followed by payment processing. At that point, it activates two IT paths: it connects to Alipay and confirms that the buyer has enough money to make the purchase; and it moves to the seller's enterprise resource planning (ERP), confirms the item is in stock, and reduces the corresponding number of purchased goods. Successfully supporting sellers' ERP systems on the cloud is more difficult and meaningful than simply increasing website traffic capacity. This is because, from the time the receipt is printed to the purchase being mailed out, different independent software vendors play a key role, but customers experience a seamless service. As Yuting Zhang wrote after a long interview with me:

> Sellers putting their ERP on the cloud was evidence of their trust in cloud computing, we upped our goal for 2013 to 70–80 percent of transactions being carried out over our cloud computing platform. [By 2015, that proportion had reached 90 percent.] This is not only cost cutting for our clients; more than that, it can increase sellers' likelihood of making money. This phenomenon reflects the fact that cloud computing is revolutionary and its value to clients far exceeds just cost cutting.

On May 17, 2013, Alibaba Group's final IBM minicomputer was taken offline. This marked an important stage in the de-IOE strategy that had begun in 2009. Saying goodbye to the final minicomputer meant that even though Alibaba Group still had some Oracle databases and EMC storage, there were no longer any IBM minicomputers. On July 10, 2013, Taobao's crucially important advertising system shut down its last Oracle database, the last one used by Taobao. These two events mark an important milestone for Alibaba's technological development. At the same time, de-IOE became a hot topic in the IT industry and how to make technology selections in the cloud era became widely discussed.

Wang believes, "De-IOE results in certain technological challenges; it is not simply a case of replacing software and hardware with other existing products. Instead, it means using new internet technology and structures to replace the old IT solutions. Alibaba is not a software or hardware seller, but this type of technological ability can be provided to third parties

through cloud services, thereby completely transforming the traditional IT architecture."

Why has Alibaba been able to be a leader in this
movement? Wang believes there are three important
reasons: First, the company was committed to the
strategy once it was made. Second, it was willing to
go the whole way and take on the technological and
organizational risks. Third, and also most important,
this seemingly impossible task required a team with
a strong sense of mission to accomplish it. From its
de-IOE efforts, Alibaba's strategic investment in technology is obvious. This includes the Apsara cloud
computing platform, databases, and the YunOS operating system, among others.

From 2008 to 2009, Wang first raised the idea of
de-IOE when working on the whole group's annual
technology budget. Every year, Wang was responsible
for modeling the company's technology budget, and
when he saw Alibaba's computing demand was growing exponentially, while core businesses were not, he
realized that without cloud computing, the whole
group's long-term growth would be jeopardized.

Making a budget is not just about money, it is about
considering the group's growth path. De-IOE was not
something that one person could decide. You would
need to analyze the situation of the group's businesses
and see which ones would be best to try out de-IOE.
Alibaba was rather fortunate because the Taobao tech
team were interested in innovating and were the first
to take on the technological and commercial risks of
the de-IOE process. De-IOE is a process of technological development, and to that end, Alibaba formed
teams, mainly from Taobao, working on the MySQL

database and its own indigenous Oceanbase database. The teams' members now come from various different departments and divisions, which is a sign of the extremely large long-term investment, which is expected to last at least five years. When the last IBM minicomputer was taken offline, the whole team was very proud.

Wang said, "Back when we were considering de-IOE, we were not only concerned with costs, which are just the most visible thing. The most important consideration was meeting the long-term development needs of the company. In the internet era, each enterprise must embrace the internet. Traditional IT hardware and software can no longer fulfill the needs of companies. Alibaba's success shows you no longer need to rely on IBM, Oracle, and EMC but can build your own more reasonable system. This means removing the obstacles companies face when putting their IT on the cloud and helping them embrace cloud computing and the internet.

"If you understand enough about the changes cloud computing may bring, you will understand that de-IOE is not just a one-time technological upgrade, it is an epochal change. Cloud computing presents a challenge to both IT and open-source software. This is a transformation in the industry, but is not simply a technology strategy decision. Because this is an IT infrastructure revolution, de-IOE is not something that every company should adopt for all of their hardware and software. I believe most enterprises have no need or ability for de-IOE. In fact, carrying out de-IOE means facing major technological challenges, market limitations, opportunity costs, and human resources issues. Therefore, the best way to help the majority of traditional companies to do de-IOE is cloud computing, which provides computing services without the IOE technologies and software. This provides a path for companies to realize sustainable development."

Cloud computing and big data bring tremendous changes to the whole industry. We need to seize this opportunity to make leapfrog advances in technology. IOE represents the era of buying computers, while the cloud signals an era of pay-to-use computing. De-IOE's best approach is cloud computing rather than buying new equipment to replace the old. This is an industry shift, not a strategic decision. The challenge lies in whether you can accept the end of the so-called private cloud and transform your faith in traditional hardware and software providers into trust in the cloud. In essence, de-IOE is a hallmark of a company transitioning from a traditional enterprise into an internet company.

On Singles' Day 2013, around 70 percent of transactions were processed over the cloud as planned, while in 2012 only 20 percent were. This is not just an increase in transaction volume. It shows that computers are being replaced by computing.

Chapter 5

The Mobile Internet Is the Internet of Being Online

The Clamor of Mobile and the Silence of Being Online

The true impact of the mobile internet is in allowing people to be online almost all the time. The mobile internet deepens our ability to be online far more than the traditional internet.

The term *mobile* appeared a long time ago, but it originally referred to something easy to carry around, not something that is online. When I flipped through a stack of old magazines I had, I discovered there was a magazine called *Mobile Computing* published in the early 1990s. So you could say the term *mobile* is already quite old. The mobile internet goes by another name, too: the wireless internet. Wireless is just a type of technology often confused with mobile, but in reality, both terms are used to describe the ability of computing to be transported with ease. We can't have mobility without wireless technology.

Wireless technology allows mobile equipment to become much more useful and makes it possible for mobile computing equipment to always be online. Wi-Fi and 3G, 4G, and in the near future 5G communication networks let us be online at any time. Looking back, the portable computers of ten-plus years ago were nothing more than cell phones that couldn't make phone calls. They were pure mobile technology, but more importantly, they were just smaller and slower PCs since they couldn't go online.

Even though Skype was free, it didn't defeat landline telephones—that was done by cell phones, which carry a fee. Free, therefore, is not the key. DVDs defeated VHSs and LPs, and in 2008 Blu-ray Disc knocked out HD-DVD to become the standard. But before Blu-ray Discs had completely replaced DVDs, online video streaming came out and displaced both Blu-ray Discs and DVDs. Now Netflix is far more impactful than Sony was when its Blu-ray won the format war. It doesn't matter how much capacity your disc has or if it becomes the standard product, you won't be able to forestall the future—and the future is being online.

In April 2000, Compaq released its first mobile device—iPAQ, which used Windows CE's operating system and the Intel StrongArm microprocessor and had a color display. It could do word processing, spreadsheets, and PowerPoint. In addition to Compaq's handheld computer, HP, Casio, and Palm all released their own personal data assistants (PDA). When PDAs were popular, everyone was talking about the threat they posed to personal computers because of their convenience. But in the end, there was no revolution, and PDAs turned out to be much ado about nothing.

Though PDAs didn't defeat the PC, laptops eventually replaced most PC usage. When we look at true industry shifts and revolutions, the size of the products isn't the only determinative factor. An offline PDA is small in size but can't compete with a laptop. Mobile is just like digitization; it is the foundation of a major change, and nonmobile equipment can't be part of the being online conversation. But when people look at mobile technology, they need to be clear about one thing: only when the mobile technology is online can it be revolutionary. Some people say, because the mobile equipment user experience is better than that of PCs, it represents a large threat, but this is an illusion. What is really going on here? The difference between iPod Touch and iPhone, as well as the difference between iOS and Symbian, can tell us something about this.

The iPod Touch and iPhone are almost identical in their functions, with the exception of telecommunications. But the iPhone exploded and the iPod Touch did not, because the iTouch could only go online with Wi-Fi and couldn't truly be online at any time.

Making phone calls and going on the internet were not inherently closely tied. Over the course of the development of the telecom industry, telephone service came first. The early large, clumsy, and bulky cell phones that could use an analog signal were replaced by digitized models that could also send text messages. Texts were the first form of digital services and were a first-generation and the most successful wireless business. Limited to 140 characters per message, texting had the highest unit cost of any data service, yet it was wildly popular and the speed and rate of penetration were shocking. The next step in terms of a data service was Web surfing, but because 2G was so slow, it wasn't until 3G that people could have a complete Web-surfing experience on digital cell phones.

Traditional cell phones, what were called feature phones in the industry, had a few more functions than the original basic cell phones and were eventually replaced by smartphones. The main difference between smartphones and feature phones is that smartphones have operating systems on par with personal computers, like Symbian, Windows Mobile, and iOS. In addition to being able to make phone calls, smartphones possessed the same functions PDAs or PCs had.

In the context of feature phones and Apple smartphones, two other companies are worth mentioning: Nokia and Microsoft. These two companies used to make smartphones. At the height of the PC era, they were the market leaders, but by the internet era, their products could only be considered half-hearted projects. In 2014, Nokia was bought out by Microsoft, and only time will tell whether they can make a true smartphone. In my opinion, the difference between Symbian (and Windows mobile) and iOS is in their online experience. The first time I had a true online experience on a phone was with an iPhone.

Cell phone operating systems will provide excellent user experiences as soon as they are fully online. Apple didn't just kill the traditional cell phone, it ushered in a new era. In 2007, when Steve Jobs presented the first iPhone, every traditional cell phone manufacturer realized that it had no future anymore; smartphones were the only way forward. The iPhone was not the first smartphone, but it was the first smartphone to get right the mobile online user experience. Not only that, the iPhone used 3G technology.

Today, the iPhone and iPad are everywhere, but very few people remember that in 1996, Nokia released the 9000 Communicator, a handheld device that could go online; in 2001, Japan's NTT DoCoMo first provided 3G mobile telephone services.

The iPhone not only defeated the feature phone, its games revolutionized the traditional gaming device industry because buying video game cartridges and disks was thinking rooted in the pre-internet past. Being online is the way of the present. Buying physical products offline is not enough to satisfy the demands of users. That's why we need the App Store. The iPhone revolutionized the digital camera industry because people take photos not just to put them up in their homes but also to share them online. The iPhone also revolutionized the camcorder because people were now able to use social media to share their videos, which were even more popular than still photos, as evidenced by the fact that in 2013, both Facebook and Twitter released short video apps. Now filming and sharing microvideos have taken even more advantage of technological advancements. The popularity of TikTok in the United States is the most recent example.

So you can imagine how something that was offline can bring about so many opportunities for innovation once it goes online. And you can see how that can revolutionize every industry.

What I mean by being online is not just going on the internet.

Your cell phone certainly has 3G/4G Web and Wi-Fi capabilities. What is the difference between these two? Could it be that Wi-Fi is faster and cheaper? The true importance is that 3G and 4G, not Wi-Fi, are accelerating the growth of the mobile internet.

The biggest difference is that 3G/4G requires only a SIM card, and all of your authentications can happen on the card. That is a good user experience. 3G/4G doesn't require a conspicuous piece of equipment or a particular action to connect to the internet. But when you use Wi-Fi, you have to use a password. In the internet industry, there is a common saying: With every extra step, you'll lose half of your users. In other words, even if every corner of your city has Wi-Fi hotspots, as long as you still have to input a password, you'll still lose a large number of users and won't be quite able to promote access to the internet.

At the same time, 3G/4G coverage is something not achievable with Wi-Fi. Trying to get 3G/4G coverage with the existing Wi-Fi technology is impossible in practice, hence the anecdote about Tibet I mentioned in the first chapter. 3G/4G and smartphones have together brought us closer to the online experience and more likely to incorporate being online into our normal everyday behavior without us noticing it.

In 2011, I observed, "What is the mobile internet? It is smartphones and 3G networks, plus something equivalent to the traditional internet, which we do not yet know about." It doesn't matter how you define the mobile internet, 3G/4G networks and smartphones are the two most basic physical ingredients. At that time, 3G/4G hadn't yet become widespread, and iPhones hadn't yet become widely used. Finding a Wi-Fi hotspot used to be a phenomenon in major cities in China; not anymore with the wide deployment of 4G today.

Today, using a cell phone to get online is easier than finding a place to plug in. This is the trend of popularization of being online. Only with 3G/4G could you experience the feeling of always being online. Always online is a topic that has been discussed for many years, and now it has finally become a reality. 3G/4G was just the beginning. 5G and other new technologies will make being online even faster and smoother.

3G/4G let smartphone hardware go online, while the iPhone's growing sales deepened our ability to be online. The biggest contribution of Apple cell phones at first was allowing users to browse web pages. Previously, this experience was difficult to realize on small devices.

The earliest iPhones' software store was incomplete, and there weren't many apps. When the iPhone was first launched in 2007, Steve Jobs didn't even mention the App Store. The most commonly used app was Safari, which enables users to browse online content. Gradually, online content was implanted right into the phone via the App Store.

I believe the App Store itself is a system that allows for online software distribution via the internet. There had never been a distribution system with such widespread coverage in the history of software.

In the past, there were many ways to get software, like downloading shareware or free software. Unfortunately, Microsoft ruined this situation because they didn't incorporate free software or shareware into Windows. I remember when Windows 95 was just released and there was a website called www.windows95.com that was the first and largest Windows 95/NT software distribution website. It was founded by Steve Jenkins in 1995 and managed software downloads and distribution, but its income was based on donations. Later, as its product range expanded, the website name changed to winfiles.com.

Winfiles.com in the end never became an online software store. In 1998 it already had over 2GB worth of Windows 95 and Windows 98 applications. 2GB at that time was a massive amount, and Jenkins even attempted to use CDs to store the software. In 1999, CNET, the tech news and info website, bought out the site for $115 million. Microsoft's Internet Explorer defeated Netscape, but it didn't establish its own application store or provide a convenient way to buy services online. Do you know how bad the experience of uploading and downloading software on websites used to be? You would run into so many obstacles preventing you from downloading an application. It was a terrible experience. In the traditional software ecosystem, software could only be downloaded online, but they never realized true online distribution. Even though Jenkins worked as product manager for Internet Explorer, Microsoft still missed out on creating its own version of the App Store. This was an error Microsoft didn't resolve until Windows 8 was released in 2013.

Using commercial CDs to distribute software is totally different from going online and looking for software to download yourself. But the App Store is kind of software distribution. Only when you are distributing like a wholesaler can you achieve economies of scale and enter a virtuous cycle. To my understanding, the App Store is a first in putting software distribution online. In the past Microsoft made software distribution offline through OEM (original equipment manufacturers) with either preinstallation or software CDs. In fact, software distributed online is not actually traditional software anymore. Most cell phone apps don't even have web addresses. In the past you had to open a uniform resource locator (URL) address to browse web pages on the internet, but now you open an app. The logic behind these apps, however, is the same as a complete website.

Online and offline software have fundamental differences. In the past, software such as Office and Windows didn't have the concept of "daily operation." As long as they were installed, you could use them completely on your own PC. But online software that may be used by only a few dozen users today might be used by hundreds of thousands of users tomorrow. The computing demands of software operation will be very different on a daily basis, and there will be a need for the support of cloud services to make the software fully functional. That is why computing must become a type of service. If computing becomes a form of infrastructure, online software will become even more widespread. Software is being redefined by being online. The old type of software is gone for good, and this is a valuable change. What is meant by a "blundering" smartphone? One that has not only failed to provide a good online experience consistent with the mobile internet but that lacks an online app system and doesn't have a good user system. Of course, I'm being a bit extreme here, but the point I'm trying to make is that all software is becoming online software as the internet turns into the mobile internet.

As I described at the beginning of this book, the cell phone QQ app hid their online function so real-time telecommunications could go completely online and you wouldn't have to worry if the person you wanted to talk with was using QQ or not. Cell phone apps have hidden URLs, thereby making websites even more online and taking away the need to enter a URL.

One day, a friend and I were talking, and he was complaining about Samsung phones: "A dialog box pops up and says it needs to be upgraded online. The box has only three options: upgrade in one hour, two hours, or three hours. I didn't want to upgrade at all, but that wasn't even an option. I clicked three hours later twice and finally gave up and upgraded the phone." I didn't ask which model phone he was using, but this is a new challenge brought on by systems being online. This is the cell phone operating system upgrade in the internet era, which is very different from a Microsoft or Nokia cell phone upgrade.

When you had to update your PC's system, it couldn't be done online. You had to buy a disk and install the update one step at a time. This means that each system update would result in the loss of many users who consider the process too much of a hassle. But cell phone software is now online. What does this mean? To be direct, whatever they want to promote, they can send it straight to you. Its operation can be done online, without disks. Before being online, this wouldn't have been possible. Today, Windows 10 does all upgrades online just like the iPhone.

Nokia used to make sure their cell phones were perfect and flawless before they shipped, but this got in the way of their ability to innovate. Now you can boldly innovate a cell phone, get it into the hands of users, and constantly update it online so it can be quickly improved. This is a major change for the industry and something traditional manufacturers didn't learn. Otherwise, Nokia executives wouldn't have made fun of Steve Jobs when the first iPhone was released, saying, "Consumers won't buy a phone whose screen can crack."

When it came to making the Windows Phone, its OS upgrades were so infrequent they were hardly noticeable. Although there already were plenty of cell phone apps at that point, they were still not operating with a completely being online mindset. The rules of the game are different in the online and offline worlds. In the online era, upgrading is no longer dependent on a disk reaching the user's hands. As long as the user is willing to use it, you will have a way of delivering the upgrade service. That is the way of doing things in the online society.

In the offline world, it's very easy to recognize a person. Everyone's appearance is different, so you can tell who they are. But the traditional identifiers like clothes and appearance are no longer available to you when you're online. You need to have a username/nickname to differentiate users.

Even traditional news portal websites are trying their best to build their account system. If they can't do it, they use other companies'. Because without accounts, news websites have no ability to provide content recommendations to users based on their likes and interests, and the value of the website drops dramatically. Account names and passwords are not inherently related. When Microsoft's Windows Phone first came out, there was no concept of accounts, just passwords, and all the password could do was unlock the screen on the phone.

Truly online accounts help you manage data. Apple made a major contribution by having accounts for its operating system from day one. If you want to activate an iPhone, you must have an Apple account or register a new one; otherwise you can't use the essential smartphone feature, the App Store.

This is revolutionary thinking. Comparing the Windows Phone and the iPhone, it's clear that the password on the former is nothing like the account on the latter.

When you open up Windows 10, the system will ask you if you want the password for the devices running the program to be the same as your online account. From this detail, you can see how important and revolutionary accounts are. Today, Windows 10 is pretty much like an online operating system and in that sense is much better than Windows Phone.

Putting an account online creates value. In the past, "using your phone to surf the internet" was a common expression. But after considering it deeply, I realized it was inaccurate in some sense. It should refer to the internet *in* your phone. What does this mean? If you find the right URL and input it, you can reach the website you want. But there has already been a massive change on the internet—for users in apps, URLs are gone and invisible. You can use 3G or 4G networks anytime, anywhere, and you have online software and online distribution systems, online applications and online accounts. Cell phones are actually permanently connected to the internet. You don't need to use your phone to log onto websites because the whole internet is connected to your phone.

Online content can appear on your phone anytime, anywhere. But it's not just on your cell phone. It's on any online device. Many people nowadays couldn't imagine a mobile internet without cell phones, but that's an inaccurate belief. Cell phones aren't as irreplaceable as they seem. The mobile internet and cell phones are now in their honeymoon phase, and it appears that the two are inseparable. In reality, watches, glasses, TVs, cars, and many more things are going online in a way that will replicate the process cell phones underwent. Cell phones are just one of many online devices, as far as a truly mature internet would be concerned. Gradually, every piece of equipment will become an internet terminal, not just cell phones. Cell phones are just the beginning.

The ultimate impact of the mobile internet is that people will spend more and more time immersed in the online world. Your time is online.

Apps Are Good, but the Web Is Great

I feel that the Amazon jungle couldn't grow in the world of apps, but it would thrive on the internet and the Web.

In its 2013 third-quarter financial report, Apple's App Store boasted some 900,000 apps. Before this, Apple announced that there had been 50 billion downloads from the App Store. In May 2013, Google Play reached 48.5 billion app downloads, not including app downloads from other Android applications stores.

As I noted above, cell phone apps overthrew traditional software and are the new driving force of this generation. But don't focus too much on how popular apps are right now. I have a feeling that cell phone apps aren't the optimal form the future of the mobile internet will take. Apps' status on the mobile internet can't be compared to that of the World Wide Web on the traditional internet.

When URLs were primarily being used, you just needed a domain name that was in accordance with local laws, and you would have a comparatively relaxed environment. The internet is a very open and fair place. But cell phone apps haven't completely followed the same path of openness and equality. Apple's iOS and Google's Android both restrict you. Even though you may have escaped the minor constraints of the URL, you're suddenly thrust into a game with new rules run by a much stricter boss, who is constantly supervising you. I believe there will be a time when you and the boss will get into a fight. I call this situation "exchanging handcuffs for leg irons." The leg irons will result in a brand-new problem. As soon as a company registers a Web address to start a new business, you have freedom, but this means you're taking on a big uncertainty, because you have to do everything from zero. High uncertainty, however, can yield high rewards. You could be google.com or facebook.com. The Web economy's scale is amazing—this is the incomparable opportunity the internet and the World Wide Web have given us.

Apps are different. Apple, the boss in this new game, has lowered the barrier to entry by giving you tools and users and even providing you with a way to receive payments. Right now, we don't know how large an economic system an app can sustain. Yet we do know that before it reaches a certain size, it will still have to rely on iOS and Android. Developing apps certainly means you can't be Apple.

But where are the limits of this new boss? I don't know, but I believe the limit must exist. If Facebook wants to get onto the App Store, its own payment system cannot be used, they have to use Apple's. Quite a few years ago, Xunfei Yudian, a voice software developed by the Chinese company Xunfei, hoped to get onto the App Store. It was rejected three times because "its features were considered too similar to Siri." After nearly one year, Xunfei deleted its voice dialing and texting and voice reminder features, and it was finally allowed onto the App Store. In essence, these events show that the once loosely regulated internet space has given way to the mobile internet, which is run by a handful of companies that set the rules of the game. The intensity of regulation from these companies will determine the level of development for the whole mobile internet. Is this a risky situation?

In 2011, I gave a speech at the Global Mobile Internet Conference entitled "Apps Are Good, but the Web Is Great." I firmly believe the Web will make a comeback and the app ecosystem will look more and more like the Web. Apple's App Store and Google Play will not necessarily continue to monopolize the market, because equality is the cornerstone of innovation.

Today, app development seems a bit like planting flowers in someone else's garden. Apple and Google have built their own gardens, and it's fine if you plant your own flowers and herbs; they'll be able to survive. But it will be challenging if you want to plant something that is bursting with life in someone else's garden. As soon as your innovation bothers the gardener, you'll have to leave if he kicks you out. You need to go to the sprawling forest; only there can you do something that will have the ability to blossom and grow. Gardens are not a friendly environment. Even though they appear clean and orderly, their ability to cultivate life isn't even close to that of the Amazon rainforest. If you want to create an ecosystem or innovate on the Internet, I believe you need a bit of courage.

When people used to mention ecosystems, they would say that they were based on a type of giving or charitable attitude. If a big company absolutely had to make an ecosystem, in many circumstances, it wouldn't create an empire instead. I believe ecosystems can only be made if everyone believes in helping each other. At this point, no one is making a ruckus about building ecosystems, because the ecosystem has already formed and it is like a tropical rainforest. In this environment, only big innovations will survive and grow. An investor once said, "I feel like in the app ecosystem, you can't grow an Amazon rainforest, but on the internet you can, on the Web you can." I believe this makes a lot of sense. My personal belief is that Facebook is the last truly successful traditional internet company. But in the mobile internet space, I have yet to see a truly successful, impactful company emerge.

It was unfortunate that WhatsApp was sold to Facebook for $19 billion. So far, it seems that the size of an app in the app space can't compare to that of a website on the internet. The natural limits of economic impact are also the main difference between an app and a website, just as the app ecosystem cannot compare with the Web. I believe apps need to move closer to the Web. This is the power of being online.

Today, apps are a significant part of the internet and work in a very similar way as the Web.

YunOS: An Operating System Created for Being Online

Making YunOS sparked a significant amount of criticism, perhaps more than I have ever received in my life. But I don't regret it.

On December 10, 2015, I delivered a keynote speech at the new release conference for YunOS 5.0 and observed: "Six years ago, when I said I wanted to make an operating system, the vast majority of people said this was crazy and I was acting like Don Quixote. Now, when we discuss how YunOS will turn into a new ecosystem and innovation platform, more people believe this is possible."

YunOS was first launched in 2011, and by 2015, according to market data from Sino Market Research, Android, iOS, and YunOS had a market share of 81.36 percent, 11.00 percent, and 7.10 percent, respectively. The data from IDC (International Data Corporation) also showed that in 2015, the total for Windows phones shipped was about 31 million, while for YunOS phones the total was about 33 million.

In China, many companies wanted to work on mobile operating systems, and some of them had a better starting position than Aliyun did. If you asked a consulting firm to evaluate our situation, Aliyun wouldn't have been considered even remotely qualified. But we were bold and diligent, and in the end, we were the only one that stayed the course and gained a substantial share of the market. This was because we were persistent and enthusiastic and believed that the project had value for society. Fertile ground stretches before us, but only perceptive farmers can produce the right crops. It was not that others didn't see this opportunity, but they just didn't have enough faith.

While people were busy discussing whether or not China should produce its own operating system, it's not really about China or other countries. I chose to believe in an OS's inherent value, and that is why we're very proud of having done it. This should be something everyone can be proud of. From day one, we always wanted to make an OS that would be different from the traditional systems. The goal has never changed, but getting there and what the end product would look like were a process of exploration.

The meaning of an operating system is different for every generation and is constantly changing. A computer science textbook definition of an OS is a program that manages the CPU, storage, and disk drives. This description is very simple and has nothing to do with networks, graphical user interface (GUI), browsers, and data.

Back in the early days of DOS and Windows 3.0, the internet wasn't yet a part of an operating system's purview; not even local area networks were. After the addition of network cards, the network systems company Novell developed NetWare, which was network connection built on top of an operating system. Novell created the Internet Packaging Exchange Agreement (IPX), which made the LAN environment centered around local servers NetWare's core competitiveness. NetWare was wildly successful all the way to the launch of Windows 95.

At the end of the 1990s, during the rise of the internet, Novell's products failed to keep up with the development of the transmission control protocol (TCP)/internet protocol (IP), and the company disappeared without a trace. The network finally became part of the Windows NT/2000 operating system. From then on, no company could rely on a business to make network software for an operating system to earn money. The fate of those types of companies was sealed for the simple reason that the network had already become part of the operating systems themselves.

Now having the internet built into an OS is considered to be the default. An OS that doesn't come with the internet would be unacceptable.

During the course of the development of PCs, graphical user interface was a key milestone, and its importance is not limited to user experience. In 1984, Steve Jobs unveiled the Macintosh. It was amazing. Unlike the Lisa workstation Apple had launched the year before, the Macintosh computer told everyone that GUI was something that operating systems couldn't do without. Steve Jobs wasn't changing the meaning of an OS based on the textbook definition. In fact, he was redefining operating systems based on user experience. In order to emphasize the GUI, Apple removed the directional pad on the keyboard and required the use of a mouse. GUI from then on became part of all operating systems. Today, it is impossible for us to imagine a device without GUI, even cell phones.

In the mid-1990s, the World Wide Web had been gradually gaining traction among the general population, and the

Netscape browser saw rapid growth, taking up a 90 percent market share. But in 1995, Microsoft's Internet Explorer (IE) 1.0 was released as a separate package of Windows 95 update. Three months after launching IE 2.0, the Web browser was available for free download. The following years witnessed the first browser war of the internet era, and in 1997, when IE 4.0 was released and was integrated into Windows, the browser finally became a part of an OS. IE 4.0 transformed the browser war, and by 2006, Netscape held less than a one-percent market share. Even if they didn't want to use IE and hated how Microsoft used monopolistic practices to squeeze out Netscape, consumers would still be unable to accept a Windows system without IE.

Operating systems and computers were linked from birth, and from the very beginning operating systems have had something called logs, which contain the most important data about the system operation. Data follows in step with operating systems. The earliest logs could only record time sharing operations. You could say that the log made sure the system was running normally. Operating systems are closely linked with data; it's just that the data was originally focused on the running of a device.

In the internet era, browsers also maintained logs. Because most of people's activity occurred through the browser, browsers used to block data that originally belonged to the OS. Thanks to the internet, even more of a person's online activity was logged on the servers. The mobile internet followed next with more and more cell phone sensor data being collected onto the OS, making the value of operating systems rise.

In the early days, when most people mentioned a mobile operating system, they were talking about having a cell phone that could make calls at any time. The Motorola operating system had no online experience or user accounts. It was essentially an offline cell phone that could make phone calls, certainly with some kind of software. At best, you could download a browser to surf the Web, but that was just a function, not an experience. In that respect, Motorola couldn't compete with Apple.

Traditional cell phone operating systems and mobile operating systems in the internet era are completely different. The traditional operating system works as the interface between the hardware system and the user. YunOS treats the internet as a separate online system parallel to the hardware system. The core of YunOS is a "being online OS," as I would like to call it. The conception of YunOS was controversial from the very beginning.

When we started working on YunOS, we decided to make a mobile internet OS, not a cell phone OS. Its online experience was built in, and the most basic part of the online experience was the account. This absolutely was a turning point for me. I realized that accounts and payment were the most basic parts of YunOS as a mobile operating system. This was the first time I truly understood what YunOS was. iPhones and Google had accounts. It wasn't just that I was learning from them; it was that I was recognizing the fact that accounts were not a simple feature but rather a very important core component. It seems trivial today, but it was hard for me to grasp at the time.

Accounts actually provide a type of online experience.

Google has a product package, called Google Mobile Services (GMS). If you want to know if an Android phone is Google-certified or not, see if it has Google Search, Google Chrome, YouTube, and the Google Play Store. If it doesn't, then it's not Google-certified. Android is telling everyone that the being online experience is part of its OS. In that respect, GMS is the same as Internet Explorer.

People who disagreed with me often said that GMS was just part of the Android application package, and therefore it wasn't the core part of its OS. That's something Google has been very clever about. When they release GMS, it is as an app. Whereas, I believe that even though GMS looks like an Android application package, it is actually also part of the OS. Again, it's just like the early days of Internet Explorer.

Further, the functions of a traditional operating system, like making phone calls, aren't core components anymore. When I started working on YunOS, everyone around me asked: "What is its killer app?" The term "killer app" is computer industry jargon for an application program that has broad usage and has never existed before. If you want to use the killer app, you need to buy the system that runs the program. The browser and its connection to internet servers were the killer app of Windows in the 1990s.

Three mobile operating systems that people talk about are all made in the United States. They are Apple's iOS, Google's Android, and Microsoft's Windows. The United States has only 330 million people, yet it has that many operating systems, and that's not including all the other ones that failed.

What's wrong with making one that might fail? I've never thought that failing was a terrible thing. The thing I hate most is raising the white flag before the battle has begun. China already has world-class internet companies, like Alibaba and others. Under those circumstances, I believed it was wrong that no domestic Chinese company had developed its own mobile internet operating system. That omission would cause China to fall behind in mobile internet technology.

Cloud computing and the mobile internet are two sides of the same coin. One can't live without the other. This world needs an option beyond Apple and Google.

By March 2016, YunOS was activated on 30 million devices. Many people, probably including Google, hadn't imagined that possible. The controversy surrounding YunOS reflects different outlooks toward operating systems in the internet era and different appraisals of technological development. Diverse opinions on so difficult an issue are quite natural. But at the core of these disagreements lies a misunderstanding of the role of mobile operating systems, a misunderstanding that is similar to how, before the Macintosh, people didn't realize that GUI was an integral part of an OS. In the same vein, before Windows 95, people didn't know that browsers were part of operating systems. In the internet era, it's also not unusual for controversy to accompany the discussion of integrating service systems and the online experience into operating systems. My position is simple: it's unrealistic for there to be no Chinese internet company that has its own mobile internet OS.

Most people underestimate the implication of this position.

But I too don't know if I am underestimating the grand challenges of my position. All I can say is that I am someone who has spent a very long period of time thinking about it. If I hadn't pondered this process for as long as I have, I wouldn't have been able to persevere this long. It wasn't about weighing investment or costs, but rather the more you think about the challenge, the more you think you can do it. By the same token, the more time you've spent thinking about the challenge, the quicker you're likely to decide whether to take it on or not. If you can't stop doing something very quickly or persevere, it means the amount of time you've spent thinking about it was insufficient. Only if you've spend enough time thinking about something can your perseverance be clearheaded.

When you're working on something, most people will gladly estimate its difficulty. Many say that making an OS is very hard, but if you don't do something just because it's hard, that means it is not important to you. If you're thinking about doing something, you must assume it will be difficult. That goes without saying. In the beginning, we had some colleagues leave, not because they weren't willing to invest the work or because it was too hard—these people would have been willing to work twenty-four hours a day. Rather, it was because they didn't have hope. You can't blame them for that.

Making YunOS made me the object of considerable criticism. I was attacked more than I had been ever before in my life. But I have no regrets.

On the surface, what we were doing was related to cell phones, but in essence, we believed in something different. In early 2011, many employees came by and said: What Xiaomi—the giant home electronics multinational that was founded in 2010 and makes smartphones, mobile apps, and so forth—has worked so hard to accomplish should also be the focus of our efforts. At the time, many of my colleagues were at a loss for what to do. The rapid growth and success of Xiaomi caused them to lose focus. If you make the mistake of supposing that Xiaomi is the reason for the existence of every cell phone company, then YunOS necessarily loses its social and industry value. However, this wasn't a problem of competition between corporations, but rather one of identifying the value of something. Our goal was to push a new internet era and transform an industry, not simply increase cell phone market share. These are two totally different concepts and pursuits. The challenges Xiaomi faces today, only they themselves know.

On June 10, 2017, YunOS 6.0 was released at the YunOS developer conference in Shanghai. It was a milestone, the last major release of YunOS. Later, YunOS gave up the cell phone market completely. Innovation is never easy.

It's Not the Internet of Things, It's the Internet of All Things Under the Sun

In the past, the process was about digitizing offline things; but today's O2O (offline to online) is about making offline things online.

My work at Aliyun allowed me to think about cloud computing, the mobile internet, and big data all together.

O2O came into wide use thanks to group buying websites.

In 2008, Groupon appeared. This website provided customers with discounted products based on service and location. Groupon was not a pure e-commerce website; it was a combination of e-commerce, internet advertising, and offline services. The group buying craze swept the globe and, at its peak, China alone boasted some 1,000 companies in this space. Almost every third-tier city had its own group-buying site. Groupon was named the fastest-growing company in history by *Forbes* magazine. People started to discover that beyond search, games, and portals, O2O—meaning systems for inducing users online to make purchases from physical businesses offline—offered a mysterious appeal.

By 2016, people gradually understood that the O2O industry that had absorbed countless amounts of capital was just a flash in the pan. But this flash helped me see that the line between offline and online was blurring to the point that O2O had become O^2. The two worlds had blended into each other. When I was shopping in the world's largest home electronics retail store, Best Buy, I watched someone look through things, and when he saw what he wanted, he stopped, pulled out his cell phone, and bought the item on Best Buy's website. This scene is very important. He didn't go home and buy something online; he bought it right on the spot. Was this mobile e-commerce or offline retail?

Suppose I run into you on the road and I say "hi." Neither of us has our business card with us, so we get out our cell phones and scan each other's QR codes. Is this mobile social networking or a face-to-face meeting?

In China, there will occasionally be regional power outages. When an area's supply of electricity can't meet the demand, the state grid has to shut down the supply, or it will affect electricity transmission across the whole grid. The commonly used method is to switch off the power. This is a direct means and can immediately shut down power to your house and your apartment complex. But the state grid has also discussed a different way of shutting off power. If every air conditioner were online, there would be no need to flip a switch. So long as you could change the temperature settings on all the air conditioners, power use would go down and there would be no need to cut off the electricity. There would be such a slight change in your home's temperature, you wouldn't even notice it.

The shade in passenger windows on Boeing 747s can be adjusted depending on passenger preference. This is a completely offline product. But the 787 isn't designed this way. The glass in its windows changes colors at the push of a button to achieve a closed effect digitally. Eventually, this kind of window could go online.

Compared with any other locomotive, bullet trains are much more online. Where the train is and how fast it's going can be seen at all times.

Hangzhou's buses have been fitted with GPS, so when you're waiting at a bus stop, you can see a display with a digital readout saying things like "Bus K10, 800 meters away." Although data won't change how long you'll have to wait for the bus, at least if you're waiting on a blazing-hot day, you won't be as anxious.

All of these examples show that the internet is making things that seemed unable to go online become online. Some people would call these things part of the internet of things, which I prefer to think of as the internet of being online and others refer to as the internet of everything (IoE). Future societies will no longer have offline things.

Many things you could never have imagined online will be able go online. Every carton of milk could be fitted with an radio-frequency identification (RFID) chip. When you buy milk, one scan would tell you where the milk came from and when it was produced and provide directions on how to use it. Smart refrigerators could calculate the amount of milk you use in a week and how much milk you have left. Your refrigerator could send you a reminder to buy more when you're about to run out. If this weren't enough, when you throw the milk in the trash, the RFID would sense this, making your trash go online.

Going back to the air conditioner anecdote, if you adopt a truly being online mentality, you stop thinking about using the remote control. What you truly want is an A/C that's totally online, that can adjust the room's temperature and humidity in real time based on your location and the changes in the local weather.

Microsoft once produced a flight-simulator game that would display a highly realistic flight on your computer. On September 11, 2001, terrorists hijacked four passenger planes and flew them into the World Trade Center and the Pentagon. After 9/11, the Microsoft Flight Simulator came under fire. Some reports alleged that the terrorists used this game for practice and simulated flying into buildings. From this, you can image how lifelike it was at the time. More than ten years later, when you use this game to land in New York, not only will it look very realistic, but you can also check meteorological data for real-life wind speeds and precipitation information. The real world's weather data gets incorporated into the game without your even noticing.

The borders of your life are blurring. This is a different kind of virtual reality. Virtual is getting real.

In 2008, the number of devices online exceeded the number of people online. The internet had more new members, including cars.

In July 2014, Alibaba and Shanghai Automotive signed an agreement to welcome a new member—the internet car. This is not just a car that's connected to the internet; it's one that drives on the internet. By driving a car on the internet, we view the internet as a new infrastructure for the car, similar to a highway. Eventually, you won't be able to drive a car without the internet. The engine that actually drives this internet car is the YunOS for cars, an operating system. Operating systems from this moment on became a core part of cars. The internet car's advent is extremely meaningful. It's a milestone, marking the shift from the internet, passing through the mobile internet, and transforming into the internet of everything.

On July 6, 2016, Shanghai Automotive released the Roewe RX5 SUV, a new electric car with YunOS for cars. SUV RX5 was the top-selling car that year.

Further integrating cell phones into cars is not the goal of the internet car, nor it is to create the internet of vehicles. The primary goal of the internet car is to liberate cars from the control of cell phones, preventing phones from becoming a barrier between cars and the internet. Eventually, this will make the phone useless in a car.

The term *internet car* emphasizes that the internet, the most important infrastructure in human history, is now as much the infrastructure of the car as the highway is. This completely changes car infrastructure. It is the most important change in automotive infrastructure since highways. From now on, internet cars will be differentiated by the unique characteristic that they are online. When they're driving down the road, they're also driving down the internet's information superhighway. This will result in an essential change in both form and content. This type of transformation comes from the computing engine that powers the car.

In the past, cars expended humanity's most important natural resources, like oil. But when cars have computing engines, they won't just become smart, they'll become smarter the more you drive them, just as humans become smarter over time. And for that, what needs to be provided to the computing engine is data, including that data the internet car itself generates. This new car will better apply data to improve the user experience, advance automotive design, and even increase the capacity of the transportation infrastructure. The internet car will take advantage of the true value of data.

In the past, when people spoke of the internet, they would say that everything had changed with the arrival of the internet, but not many seriously considered how the PC had laid the foundation for this. When people talk about the mobile internet, they're often overlooking how cell phones have played an integral part in its widespread adoption and use.

Cars will be like cell phones and PCs in the history of the development of the internet. Without PCs, there would be no Web-centered traditional internet. Without cell phones, there would be no mobile internet. Cars will play a similar role with regard to the internet of everything. Its impact on people and industries will be far greater than that of any other piece of equipment.

In the traditional internet era, people focused on the browser more than the PC. It might make some sense to say that the internet is very distant from manufacturing. But, in the internet of everything era, if the internet isn't tightly linked with manufacturing and equipment is ignored, the results could be fatal. The relationship between traditional industries and the internet industry will be redefined by the internet of everything, which presents us with a massive opportunity. Grasping this opportunity will mean there is no boundary between the internet and traditional manufacturing industries.

There is an Xbox 360 motherboard sitting on a bookshelf in my office. The early era Xbox lay flat, but to stay in fashion, the model 360 had a sleek body and was placed upright. However, then there was the problem of overheating, which caused a tin solder to soften and drop the chip, and a red light flashed endlessly. This problem cost Microsoft $1 billion. So no one should overlook hardware and think that it's so simple mastering horizontal placement means you can change it to vertical with no problem. Apple making both cell phones and an operating system is of course good; and it's not that we don't want to learn from Apple, but that we don't have the ability. Just because something is logical doesn't mean it's possible. Being logical is useless. Many of my coworkers believe that if something makes sense, then it must be the right thing to do, but logical doesn't equal practicable.

Making good hardware is hard. Making hardware online is even harder.

Chapter 6

Pursuing the Dream of the Online Cloud

Computing is the power of humankind, and data is the wealth of the world. That's the dream of the online cloud we are pursuing.

Making Computing Online

The being online world has a new value system, and many unpredictable things have happened with it.

There is no longer a difference between large and small, manufacturing companies and internet companies, traditional internet companies and mobile internet companies. The lines between many things have blurred or even disappeared.

When it comes to cloud computing, most people are either confused or they believe it's something far removed from their daily lives. But the cloud is right beside us.

In the beginning of 2013, I received a letter from a railroad worker who works in a very remote city called Ankang in the west of China. His name is Lei Wu, and he listed a lot of problems with Aliyun services in his letter. This was the first I'd had heard of Ankang in Shanxi province. In October 2013, to show my thanks for his letter, I invited Wu to the Aliyun Developers Conference in Hangzhou.

Lei Wu's mother liked to watch police procedural dramas, and she was worried that Aliyun was a kind of pyramid scheme and would try to trick her son, so she wouldn't give him permission to go. Lei Wu left without his family knowing, but he too was worried about being conned so he rejected the offers of Aliyun employees to pick him up at the train station and instead took a cab. I was told later that he thought this would make it easier for him to escape in case anything went wrong.

I was very interested to learn from him how a railroad worker had decided to use cloud computing. When I met Lei Wu, I excitedly told him that I'd love to go to Ankang to visit him. After that, Lei Wu would frequently remind me about my promise, but because I was always too busy, I couldn't make time to go. Then one day, he told me, "Dr. Wang, if you don't come to Ankang, I'll be in big trouble." I didn't realize until I got to there that Wu's family, friends, coworkers, and boss all refused to believe him when he said a cloud computing researcher would come to his mountain town. If I didn't go to Ankang, people would see the photo Lei Wu and I had taken together as evidence of his lies.

With this type of story, you don't know whether to laugh or cry, but when I reached Ankang, I couldn't help but respect Lei Wu. The locomotive depot he worked for was responsible for maintaining a thousand kilometers of train track in the middle of the Qinling Mountains. Railway maintenance workers have to spend the whole year doing track work in the field to guarantee train safety. Lei Wu, who hadn't even graduated from high school, used cloud computing to send urgent documents and announcements to his colleagues working out on the thousand-kilometer stretch of rails he managed. He would send out train delay and other safety notices in order to keep the trains safe. When some area experienced a collapse or other hazardous condition, railroad workers were able to take photos with their cell phones and quickly share their information using the cloud, so that the people in charge of other sections would know as soon as possible. Limitations of space and time that had originally been insurmountable were now easily overcome thanks to cloud computing.

When Lei Wu showed me that he was able to access all the documents in the cloud in a thirty-kilometer-long tunnel under the Qinling Mountains, I realized what a boon cloud computing was and how it had become an integral part of Ankang's railroad safety.

What left a deeper impression on me, however, was what happened at a small station in the Bashan Mountains, where I saw a little boy and girl sharing a cell phone. The train station was the only place in the whole area with Wi-Fi. This scene moved me indescribably and was the internet equivalent of an ancient Chinese story about boring a hole in your wall to use the light from your neighbor's house. The "light" allowed these children, who hadn't yet left their mountain village, to understand the outside world and connect their dreams with the world beyond.

I would like for cloud computing to be able to serve more people like this and help realize every Chinese person's dream.

I'm also grateful to Aliyun's other users, particularly ones I haven't met yet. It is these people whose creativity on the cloud and the internet constantly inspires me, who make me think more deeply and discover things that are happening all around me. Without Aliyun, I would never have met these people. The internet changed Lei Wu and me. It turned many more people into dream chasers, including Aliyun itself.

Breaking through the Fog Surrounding Cloud Computing

Back to the second half of 2011, Jiang Liu, then a reporter for Chinese Software Developer Network (CDSN), reported on Aliyun several times and wrote down his impressions about a company pursuing the dream of putting computing online.

Compared with Taobao and Alipay and other companies under the Alibaba banner, Aliyun, founded on September 10, 2009, is more low-key. Perhaps that is because the company was launched at the same time Alibaba Group was celebrating its tenth anniversary. Amid all the pageantry and celebrations, most observers either did not notice the birth of this new company, or they thought it was just a part of Alibaba Software.

From 2009 onward, aside from large-scale hiring of tech personnel at the end of that year, Aliyun practically disappeared from public view. Even though in 2010 cloud computing had already become a major focus of the industry, there appeared to be no major moves from Aliyun. An announcement of products that had been set for the end of 2010 was pushed back to some unknown date in 2011. But around this time, Aliyun's tech layout was revealed — distributed computing systems, distributed databases, a mobile operating system, a browser, an input method, search, maps, email, translation. All of this triggered heated debate in tech circles. Rumors circulated around Aliyun concerning everything from extremely high investment and R&D difficulties, to a tech team that seemed too good to be true and the imminent danger of failure. Even worse were some former employees who claimed that Aliyun was "unreliable and full of crap."

At the developers conference of October 24, 2011, Aliyun unveiled for the first time the elastic computing platform, open data processing services, an app development platform, a cloud app platform, open storage services and CDN, open structured data services, and a full line of other products.

But many questions still surrounded Aliyun. Why should an e-commerce company also work on cloud computing and a mobile operating system? Why have so many product offerings from the very beginning? How could you make a profit from cloud computing services in China? What exactly did all the elite tech staff assembled at Aliyun do over the course of three years? With these questions in mind, I flew to Hangzhou to meet with Jian Wang, head of Aliyun.

Jian Wang, known at Alibaba as "Doctor" because of his PhD, is a technology legend. A Hangzhou native, Wang is very tall, has the bearing of an academic, and speaks clearly and quickly, with a sonorous voice. Although he did not have outstanding grades growing up, he went on to earn a PhD and develop a unique way of thinking.

A dream of a technology platform
Aliyun's office, located in Hangzhou's Wen'er Road West Lake International Science and Technology Building, is very large, full of activity, and everywhere the eye can see there is an energetic technology staff. The company has 1,200 employees, 80 percent to 90 percent of whom are engineers, including China's first ACM programming competition champion, Chenxi

Lin. The core technologists are split into four groups: the basic platform, basic services, cloud computing business, and YunOS. The company has no sales department.

Google is the world leader in cloud computing technology, while Amazon is the best in cloud computing business. AWS [Amazon Web Services] has already become the mainstream cloud computing service, used by Netflix, DropBox, Zynga, and other major corporations. With several hundred million dollars in income, AWS still has great potential for further growth. But in China, the market is still up for grabs. The battle between platforms will determine who comes out on top. This goes without saying.

Alibaba is the closest thing China has to Amazon, and in e-commerce, the Chinese firm may even be more comprehensive. Alibaba has considerable resources and strengths, and something outsiders do not know is that it also has leading edge technological capacity. Therefore, it is not strange that Alibaba would go into cloud computing and join in on the platform war. Nevertheless, Alibaba's boldness is impressive. Jian Wang said that what they are trying to do could be interpreted as Amazon + Google, and eventually going beyond even that. Amazon's major contribution has been to make it easy and effective to access and utilize computing, while Google has massively scaled up the cloud computing technology. Aliyun wants to accomplish both of these things by adopting the Google model for technology and the Amazon model for operations. Meanwhile, they are also working hard on one other thing—data. They want to provide data computing power to everyone at full scale. In other words, you could make a search engine for the whole internet and you would no longer be limited by data computing power.

Aliyun's mission is expressed differently than at the definition of cloud industry—build an advanced data-centric cloud computing platform. Jian Wang said proudly that three years after the mission statement was first articulated, not a single word needs to be changed. Their understanding and evaluation of cloud computing has been sound from the beginning. Why do they emphasize data so much? Jian Wang believes data will become an incredibly important strategic resource for countries and corporations. Furthermore, a revolution has occurred in the last few years—data can all be gathered online. Moving from offline to online, data now has strategic value. Evidence for Wang's belief can be found in the current global craze around big data. International and Chinese observers have said that data will become the new oil.

Cloud computing plays a critical role in making data valuable. Computing power is a part of national and corporate core competitiveness. Google is a good example. They took data that everyone could access and turned it into their own, then they built an enormous company on this foundation. Similarly, oil 6,000 meters below the ocean surface is in theory everyone's, but other people don't have the ability to access it. Only Google has the technology to unearth this data's true value. Aliyun's goal is to become a company with the greatest computing capacity (not the most servers) for anyone who need computing power to do their own job. This is why platform matters.

Transcending Hadoop

Wang mentioned multiple times that Aliyun's biggest differentiator is its determination to develop its own competitive core technology. And after three years of difficult work on technologies that required restarting projects from scratch, they finally saw success and produced Apsara, a general large-scale distributed computing system. It is now able to support more programming models besides classical MapReduce. Further, Aliyun's tech team was able to build email, search, map services, elastic computing, data processing and many other functions on the Apsara platform. In other words, search and email are on the same ground floor. Google is the only other company in the world to have accomplished this feat.

At this point, Wang was somewhat moved. "This involved many specific technology problems, some of which were extremely challenging. For example, the scheduling system of Apsara required a considerable amount of struggle and hard work from our development team. But if you don't operate at this level of detail, you can't know how hard this was. In the beginning, many people did not believe we could do this. But we succeeded, and that is something we all have good reason to take pride in."

Chinese companies rarely decide to take the big risk of working on computing platform development (like an operation system for the cloud) by themselves. How did Aliyun decide to do this at the time? Wouldn't it be impossible to explain away if the result was failure?

Wang responded by saying that you can't avoid confronting the major battles head-on when you are building a computing platform. If you try for guerrilla warfare at the applications level, you might win some battles, but you can't win the war. Fortunately, the conditions were good for the development team: Alibaba Group has been patient and supportive, with Jack Ma one of the effort's most visible cheerleaders. The Group's CFO said that money was no issue, and many employees and users were also very supportive. Wang said, "Thanks to this favorable environment, even if we had failed, you could only say that we didn't have the ability to do it." After the success of Apsara, Aliyun finally cleared away the doubts that had been surrounding it, and the team quickly came out with many other products running on Apsara.

I then asked another question that I had prepared in advance: Why not simply use Hadoop? Wang said: "Everyone thought if we just used Hadoop, it would be better and faster. Hadoop of course is very valuable for processing offline big data, but it could not solve our problems of launching public cloud computing services, because the cloud services we had online had already exceeded the abilities of Hadoop. This was related to our company's positioning. Today, Apsara supports all of Aliyun's many different services, including big data processing. From this perspective, you could say Apsara has already transcended Hadoop."

He added that now there are many companies that call themselves cloud computing firms, but because the technology barrier is so high, very few companies actually have the technology wherewithal to do true cloud computing. Some companies set up 100 servers and say they are doing cloud storage and the so-called private cloud. They say that if they scale up individual corporate solutions, they are cloud computing, but in reality, this does not count. Additionally, general computing systems are much harder to do than vertical application systems. Although some applications may be more efficient using the latter systems, the effect of a comprehensive platform going online is completely different.

With the support of solid ground-floor technologies, Aliyun has the confidence that it has a competitive advantage in scale and cost. Wang brought up the case that he is most proud of. They use data clusters made up of 1,500 machines. While one cluster is processing data, it could also be used for a few additional tasks, and complete the graphic rendering, which would essentially be free. That's the essence of cloud computing.

Dispelling the fog

I asked, if Aliyun is focused on making a platform, why did it also work on YunOS? What is the path moving forward going to be?

Jian Wang was honest, saying that they were indeed fighting on two fronts and it was very difficult. But mobile devices are another challenge that cannot be avoided, because simply making apps could not solve all of their problems. It's an opportunity nobody wants to miss, and of all of the companies under Alibaba's umbrella, Aliyun was in the best position to make a mobile OS. Further, the mobile internet and cloud computing are inseparable. They are two sides of a coin. If you only did cloud computing for the traditional internet, it would not count as a complete success. Cloud computing does not grow out of the mobile internet, but it must blossom and bear fruit on it.

The January 2012 update of YunOS was considerably more user friendly. There also needs to be a new version every two or three months. While K-touch is already preparing more models and tablet PCs, Aliyun is in talks with a second cell phone producer. But more importantly, they are bringing the internet operations mentality to mobile phones.

I asked, how many years do you think it will take for Aliyun to become a profitable venture like Amazon?

Wang confirmed the group would continue to invest 1 billion RMB per year for ten years in Aliyun. But Aliyun might realize profits in twelve to twenty-four months. Now, the company can continue to see its income rise without hiring a single new employee.

"My number one goal is for Aliyun to provide more computing resources outside the group than within. This goal is most definitely within sight. My number two goal is to have computing scale on par with Amazon. It definitely will take less than ten years to get there."

Aliyun's strategic direction is correct. The boldness to pursue indigenous innovation is worthy of respect. You can see from the Aliyun platform's cooperation with CSDN that the product is reliable and trustworthy. Aliyun's Hangzhou office has a couplet hanging by the front desk that likely leaves an impression on all of the company's visitors: "Dreams are born from reaching for the clouds / Code achieves all things, as grains of sand form the land." Let's all wish Aliyun well.

I myself have been changed by pursing the dream of the online cloud. Otherwise, I would have likely remained a university professor. Several years after this article, I had the opportunity to speak with Liu Jiang again and told him,

"When other people had no faith in Aliyun, you believed in us." Making dreams come true is unimportant; having dreams in your heart every day is what brings you happiness. Many times, realizing dreams can be the beginning of a tragic life, because this may also mean you end up losing your goal of life. Having a dream in life isn't easy.

But the internet is like tinder-dry wood in my heart, and a flame will be lit one day.

Invaluable changes in history come from accurately grasping the demands of users and listening more closely to them. Failed revolutions, on the other hand, occur because they're too removed from the needs of the users. The value of being online is the building of a natural bridge linking users with decision makers. Users vote with their clicks and their feet. The low cost and high efficiency of being online have empowered the users themselves.

Cloud computing has no inherent value. Its value appears only when it encounters users. Below are the stories of some users. I'm telling them not because they succeeded, but because they had the boldness to take risks and seek progress. They are the benefactors of cloud computing.

Tradition and Innovation Hand in Hand; Large and Small on the Same Stage

On September 26, 2013, Tianhong Asset Management Company successfully went on the cloud, making it the first finance company to move its core settlement system onto a de-IOE cloud computing system. This move reduced its account-clearing time from eight hours to only thirty minutes, increased the number of peak-value transactions per second from five hundred to five thousand, boosted the total number of same-day transactions from 10 million per day to 300 million, and expanded the number of supportable users to 100 million.

Though the Tianhong monetary fund had previously just been a little-known domestic currency fund, it became one of the three largest monetary funds in the world following its release of Yu'e Bao on the open platform of Alipay, as reported by internet maven Mary Meeker in 2014. Yu'e Bao is a cash management tool specifically designed for internet users. It is a simple, fun-to-use mobile product built within the Alipay app, the most popular payment platform in China, with 160 million accounts and technology that enables same-day settlements. The assets under Yu'e Bao management went from zero to nearly $90 billion in under ten months, making it one of the top three global money market funds by the measurement of asset under management (AUM).

Yu'e Bao on the cloud

"Traditional financial planning products were always aimed at the top 20 percent of earners, but thanks to the internet's long-tail effect, Yu'e Bao goes after the other 80 percent," as Yisi Wu wrote in *IT Value* magazine. "This type of population weren't the movers and shakers in the traditional financial system and had trouble accessing high quality services, including even the 'little white' [a term referring to someone who has no professional skills] clients who have no concept of personal finance. Internet finance lowered the barrier to entry, and with the internet's speed, Tianhong was able to create an 'ant mover effect'—30 million clients at 3,000 RMB apiece formed a 100 billion RMB fund, the largest in China."

Yu'e Bao's launch was built on Tianhong's original data centers. In July 2014, Tianhong began preparing for Singles' Day, and based on the estimated number of transactions, it was almost guaranteed that the back-end IT system they had wouldn't be adequate. So Tianhong made the bold decision to ditch the traditional server + database + storage framework and move all of their back-end systems onto Aliyun. Chinese Software Developer Network (CSDN) once interviewed Zhenhua Fan, the head of Tianhong's innovation support department, and others on his team.

Yu'e Bao and using the internet were firsts for the fund industry. We were crossing the river by stepping from stone to stone. In order to stay safe, we used the traditional IOE framework for Phase I, investing over 4 million RMB. But we never imagined that the amount of data and transactions would increase so rapidly, far exceeding our platform's capacity to handle it, and forcing us to move on to Phase II. If we were still using the IOE model, I would estimate it would cost at least another 50 million RMB. Add to that the additional human resources required and routine maintenance in the future. This was simply not an acceptable way to complete the goal of Phase II in three months.

With no clear path or examples to follow, they were not certain how long they would have to wait or if they needed to invest tens of millions more RMB. There were not even foreign companies that they could learn from. But would the market give them enough time to wait?

Although there had been no international finance firm to use the public cloud, this was not because of technology issues. It was the result of attitude, understanding, boldness, and foresight. But the market forced us to move forward, so the decision was not as complicated as you would think.

The facts showed that "a choice without options, turned out to be the best option," Fan said.

To move software built on IOE to Aliyun required considerable development, alterations, and optimization. But on such a tight time frame (at most three months), first it was key to cultivate trust among all the collaborators.

"The internet is an integral part of Yu'e Bao. There is a large amount of data and services, and response times must be fast. On such a short time frame, communication among Tianhong, Jinzheng, Aliyun, and Alipay required immediate work. We didn't have time for market research, nor were there successful case studies to look at. You could say that the main reason we were able to work together is because of mutual trust, resource sharing, and fearlessness," Fan said.

De-IOE is easier for moving from IBM minicomputers to PC servers in terms of the hardware switch. Much more difficult is getting rid of the databases that the application had been closely integrated with, that is the Oracle databases.

After moving to the cloud, "During the trial phase, the real-time rate needed to be 3,000 TPS [3,000 user accounts could be opened per second]. But the test results showed 5,000 TPS and maxed out at 12,000 TPS. . . . If we optimized the system based on our past work experience and our previous hardware and equipment, our functionality doubled."

In addition to this, Fan is also satisfied with the security: "During the test, Aliyun's database servers shifted very quickly. Add to that our zero-tolerance policy, and hot-swapping all of our hard drives, our data was incredibly safe. To this day, we have not had a single problem.

Honestly, Aliyun is already a very safe platform, especially when dealing with high traffic flows and large influxes of data, they are much more stable and safer than many small companies' own closed systems."

They spent thirty-six hours each time they tested migrating to the cloud, a process that they repeated over and over to guarantee there were no errors. It was not until September that they were formally able to complete the switchover.

Small as a company, big in actions

Huxiu Net is a new tech media founded in May 2012. Its name means "tiger scent" in Chinese, and it is famous for its biting commentary challenging the traditional understanding of business.

In February 2013, the site suffered what has now become a well-known attack. The site published an article entitled "Real-Time Record about Our Denial of Service on February 27 and 28."

At 6:00 p.m. on February 27, Huxiu received the following abnormal alerts from four security monitoring platforms:
1. TCP link abnormal
2. CPU usage has abnormal increase
3. Overdue response time for http
4. Abnormal usage of bandwidth
5. Database pressure rising

Huxiu made the appropriate changes to the servers, but the volume of the attack was too large and had already exceeded our bandwidth.

In reality, Huxiu had already been subject to anonymous phishing and password breaking attacks.

It appeared that the hackers had been preparing for a long time.

At 6:05 p.m., as per Aliyun's procedures, the "cloud shield" was activated and began trying to ward off the attack. We received many recovery notices, and it began to look as though everything had gone back to normal. Huxiu's traffic had dropped to 1–3 percent of the attack, when the hackers' traffic volume was ten times the bandwidth Huxiu could handle.

At 7:50 p.m., the attackers began to add pressure. The attack volume reached the upper limit of the "cloud shield" (or rather the defense's free limit) and the servers suddenly collapsed.

With the help of Aliyun staff, by 9:00 p.m., Huxiu had gotten back to normal, but then the hackers once again stepped up their attack.

Huxiu Net had seen the "cloud shield" in action twice. At its worst, the attack took up forty times our bandwidth, but with support of Aliyun, the attack didn't completely take us down.

Later, the hackers changed tactics, and Huxiu went down again. This time, Aliyun's team in faraway Hangzhou used their server load balancing to defend against the attack, and once again, Huxiu was back up on its feet. After the attack had been proven ineffective, the traffic volume began to taper off.

The next day, Huxiu was the object of several different types of attacks, but each only had an impact for a short period of time. Aliyun again provided assistance in real time, and the attacks did not have a very large effect on Huxiu.

Based on this experience, we feel that there is a very big difference between defending against hackers in the cloud era versus the traditional private server era.

Several years ago when our site was attacked, if it were a small attack, we could deal with it by boosting our defense capabilities and filters. If it were bigger, we would have had to call our server department and increase bandwidth, buy more physical servers, and contact our firewall provider to get stronger filtering equipment. If the attack were more intense, we would have to raise the white flag and surrender.

Because of the physical distance between the servers and the office, this whole process took six hours in the traditional server era.

But in the cloud era, the time it takes to defend against attacks has already dropped dramatically. Small attacks result in an automatic increase in filters. And cloud companies can provide a seamless increase in temporary defense capacity. Response times and resource distribution are faster and more flexible. This means that network attacks will no longer be such a dangerous force in the future.

If we view the cloud's defense technology as a basic feature of the internet infrastructure, then technological empowerment will indeed change everything.

There is a Uyghur music site that gathers and shares Xinjiang's folk and original music. The founder, Erkin, is a music hobbyist and part-time coder from Xinjiang. After graduating from college in Beijing, he returned to Xinjiang.

When he returned home, he discovered Xinjiang was losing much of its music heritage. Many ethnic music forms had already vanished, and no one was preserving them. Eventually, he created a small website on his own, bought two virtual servers on Aliyun, and began searching everywhere for old artists' music to put on his website. As the website grew, many Xinjiang musicians sought him out in order to work with him. A few years later, it was my good fortune to invite him to visit Aliyun's developer conference at Hangzhou.

Being online is not simply about saving money. It's about breaking free from the traditional fetters of an industry and creating a completely different, more convenient service. Being online makes innovation more efficient.

Computing Becomes a New Capability of Human Beings

Cloud computing provides the best ammunition for creative people to make their dreams become a reality.

Many years ago, I attended a small gathering in Palm Beach, California. At the event, Bill Gates invited George Lucas to give a speech. Lucas said something that left an impression on me. He was talking about his experience directing *Star Wars* and said that if he had to do it over again, he wouldn't try to use new technology to make a new movie because that was incredibly painful. At the time, I thought if I had the opportunity, I would jump at the chance to try out the newest technology to enjoy the experience of doing something I had never done before.

In January 2016, I went to Las Vegas to attend CES. At the booth of Hai'er (one of the largest consumer electronics companies), I stopped for a moment. What caught my eye wasn't the curved, ultra-thin 8K OLED TV, the laser TV, or naked-eye 3D. It was a TV clip playing the animated movie *Kunta*.

China's first 3D animated feature, *Kunta* was released nationwide in China in July 2013. The movie tells the story of a child named Spinach who accompanies Kunta (a legendary hero) on his path to becoming a hero. *Kunta*'s rendering was done on Aliyun, a fact that was widely reported at the time. So, when I saw the movie being played three years later in CES, I had a sense of being transported back in time.

Kunta was the first movie to use large-scale cloud computing to support post-production rendering, which is the key final step in achieving 3D visuals for an animated movie. The amount of time the rendering of each frame takes depends on the model's complexity. The more complex the model, the more time it takes to render. It has been estimated that, if each minute requires 9,000 (60 seconds x 30 frames x 5 lays) image lays and each image lay takes an average of two hours to render with a resolution of 2K to 4K, a 160-minute-long movie would take around 2.88 million hours to render, the equivalent of operating one average server for 328 years. And for a 3D movie, you have to double that number.

The Hollywood blockbuster *The Croods* was produced by DreamWorks studio, whose CEO Jeffrey Katzenberg said, "We spent eight years on *The Croods*, six thinking about the story, and the 3D model rendering alone took over 80 million hours." Katzenberg's words are representative of the entire industry and explain the bottleneck caused by traditional graphic rendering.

Avatar used 4,000 of New Zealand's Weta Digital's servers. In 2008, the computing power of that many servers would make Weta rank somewhere between 193 and 197 on the list of world top 500 supercomputers.

Kunta director Lian Li encountered similar problems. "China long ago had the ability to make a high quality ten-minute-long cartoon, but almost no one had even attempted a ninety-minute-long feature. Before I attempted to do this, I had no idea it would be so difficult. When we were making one of the bad guys, the Lizard King, we discovered the several hundred computers we had in our workshop would not be enough to complete the rendering successfully and maintain the proper image feel and quality."

Lian Li had worked in the advertising industry for more than twenty years and poured his life savings into *Kunta*. Over the course of three years, Li's whole team was working exclusively on the animations, but computing power had turned into a major obstacle. "If you want a 3D animated movie to be as lifelike as possible, you need to make complex models—the more complex, the better. That's the only way to be detailed enough. But that means your demand for rendering ability also has to be very high."

Li had already prepared to give up on the complex graphics. Even if they were uglier, at least it would be possible to make them. Using the more refined process was far too difficult. Fortunately, Lian Li discovered cloud rendering.

In July 2011, using an Apsara server cluster of 1,500 servers, it took around forty minutes to complete the first cloud-rendered professional cartoon, with 6,654 3D images, and a resolution of 3,076 x 1,080. This was the largest rendering cluster in Asia at the time, and also the world's largest rendering system that used a general cloud computing platform. Using cloud computing for rendering allows you to split up one or many tasks into however many component parts you want and carry them out simultaneously on the computer clusters' various nodes, thereby saving a considerable amount of time.

This is why it only took three months to complete all the rendering work on *Kunta*, compared to the thirty-plus months it would have taken using the traditional method of buying servers and creating your own "rendering farm." Guaranteeing

Kunta's production schedule was a challenge, and reveals one of the main difficulties of cloud rendering. Because every day required a different number of servers, we had to make adjustments in real time according to the producers' demands. Rendering is a bit of a chaotic process. The studio comes up with the early-stage ideas and models, then the special effects team has to quickly render the models and make updates and adjustments as they watch the early versions of the movie come out. The upshot is, cloud rendering is dozens or even hundreds of times faster than the previous method.

Weta Digital hired more than nine hundred people to work on *Avatar*, and used a 4,000-server cluster that took up 10,000 square feet, 35,000 processors, and 104 terabytes of RAM to render 240,000 images.

In contrast, at its busiest the *Kunta* team used up to 6,100 Aliyun servers, with 150,000 processors, 400 terabytes of RAM and more than 10 million hours of computing. The tasks were carried out by an automatic process using Aliyun's Web-based self-served interface. Without cloud rendering, it would have been very difficult for *Kunta* to have been made on schedule. It saves rendering time for more creative work.

Even though large animation companies have their own "rendering farms," they still cannot keep up with the rapidly increasing demand for computing power. Only by giving them enough resources at low costs can you enable all creatives turn their ideas into something with commercial value. Cloud rendering can provide massive amounts of highly effective computing power in a short period of time. Taking three months to complete a project that Bocai's (*Kunta*'s production company) rendering farm would have spent forty months completing results in a 13.3-fold increase in efficiency. If they had tried to expand their rendering farm to be able to complete the job in three months, Bocai would have had to spend 28 million RMB just on hardware, not to mention the problems they would have had maintaining and recycling this equipment. Simply put, using cloud rendering resulted in Bocai reaping savings of 91 percent.

Later, a Zhejiang University student spent 200 RMB on cloud rendering to make a five-minute animated short, which earned him a design award from his school. These examples show that more and more small- and medium-sized creative companies and individuals will be able to access computing power that only large corporations could have used in the past. They will be able to survive and grow based exclusively on their own creativity. Low costs are one of the main advantages of cloud rendering. What many small-scale creative companies and individuals lack is not creativity, ideas, or stories, it is the ability to invest in hardware and get technological support.

You can imagine that if they used only one computer for rendering, one simple task would take a week or two to complete. This would be far too costly in both time and energy. Cloud rendering technology can help complete tasks at a low enough price and provide computing power on demand. As a public utility, cloud computing provides the best ammunition for creative people to turn their dreams into reality while also challenging their creativity.

Kunta was an opportunity, to quote director Lian Li, for two crazy people (Lian Li and Jian Wang) together to make something impossible a reality. The result was the successful screening of *Kunta*. I believe that cloud computing can liberate creators' productivity.

The most important thing for artists is to be able immediately to see their creations and start making edits and alterations. The more time lost, the fewer opportunities to make changes. Cloud computing provides a massive amount of power that can accomplish things in a short period of time that would previously have taken a long time. The result is that artists have more time and opportunity to try endlessly to improve their works. After all, humans are limited by time available. It's like typing. If you can type slightly faster, then you will have more time for other things. In other words, computing power is indirectly generating more creativity.

Kunta's whole production schedule was three years, including idea generation and editing. If there were no cloud computing, the rendering alone would have taken three years, and this movie would never have been made. In this respect, computing power is not just a matter of fast or slow. It is more an issue of having or not having.

Computing has become a new power, and a new type of core competitiveness.

From the story of *Kunta*'s production, you can see that a small company can now do things only a large company could do in the past, and cloud computing is giving wings to the bold entrepreneurs and creators willing to grasp this opportunity. To borrow the theme song from *Kunta*: Impossible is an impossibility.

It's my understanding that Lian Li is now working on *Kunta 2*.

Data Is a New Natural Resource, a New Source of Wealth in the World

Notarization is an important function with a long history. In the slave system at the end of the Roman Republic, there was a particular scribe position known as the *tabellio*. They had legal knowledge and could use Latin abbreviations for legal documents. They received a set compensation because of their professional status. The *tabelliones* are considered the predecessors to the modern notary profession.

After China's notary system was restored in the 1980s, the number of notary organizations in the country grew fivefold from six hundred in 1980 to three thousand in 2013. Now there are more than twenty thousand notaries, fifteen times more than three decades ago, and they give out over 10 million notarizations, 110 times as many as in 1980. The notary profession has enjoyed three decades of high-speed growth and, almost without anyone noticing, has entered the internet age. In China, when the courts hear cases, their most basic requirements for the evidence are: it must comparatively fixed and can't be manipulated; and it can be presented in court and entered into the official record.

In the online society, telephones, real-time communications apps, server codes, and many other nonprint documents need to be verified. Compared with traditional documents, the process of verifying and notarizing online electronic data is fraught with a different type of difficulty. It is not only a problem of volume. Rules and regulations have changed. Legally, there is no way to confirm this data's accuracy and completeness. If electronic order receipts, e-mails, QQ, or MSN records are simply downloaded from the internet and presented as evidence, the court will not recognize them.

It is possible to accomplish legal verification through the traditional methods, but it is very complex. One way is to record yourself talking in front of a notary in the notary public's office. But this method is obviously very inefficient. Obtaining certification, transferring the files, and keeping them saved all have related costs, not just monetary. It would be difficult if not impossible to use these methods at scale.

Ancun Technologies wants to use cloud storage to change this, starting with phone-recording authentication. Ancun's products not only record phone call content and length, they also save the caller and the recipient's phone numbers, as well as the time the phone call began to the time it ended, down to the second. If you want to record a phone call, dial 951335 and the recipient's phone number and the call will automatically be stored on secured cloud storage provided by Aliyun. You don't need to carry it with you, the data won't be lost, and there is no time limit. Anywhere there is the internet, you can listen, download, and use it for evidence. This is the most accurate and complete record of a conversation, and the system is connected with authentication and notaries, so the recordings can have legal notarization and have legal force. In November 2011, Ancun's phone call recording platform went online.

In May 2012, the Hangzhou West Lake Notary Bureau received its first application for notarization for a recording made on the Ancun system, and after careful review, the notarization of the phone record was granted the next day.

On January 1, 2013, the formal enforcement of a new law listed digital data as form of evidence, meaning it could be used as legal evidence. The internet helped created the digital evidence industry.

With the internet being as widespread as it is today, the division between netizen and consumer is fading, the boundaries of enterprises are blurring, and data has become a core resource and will deeply impact business models and even reshape corporate culture and structure. Notarization is an ancient and staid traditional industry that seems to have changed very little. But being online has breathed new life into the notary industry.

Finance is also like this.

AliFinance's website introduces the company like this: AliFinance uses an internet data-driven operation model to provide sustainable, inclusive e-commerce finance services to small and micro-companies and individual entrepreneurs who would not be able to access credit-only petty loans (small amount, short-term, and convenient repayment) through traditional channels.

From its first day of operation in October 2009, AliFinance has been using Aliyun's cloud computing platform to process data and conduct data modeling optimization in order to provide purchase loans and credit loans for the shop owner on Taobao. AliFinance highly values cloud computing technology and was an early adopter of cloud computing. When Aliyun was still a new business, they had the courage to work with us. You could say Aliyun and AliFinance depended on each other, developed together, and showed how innovative a large company can be when it encourages small internal businesses to grow together.

AliFinance's small loans are totally different from traditional banks' microloans. It does not require collateral; it is purely credit-based. AliFinance uses big data mining and does not require too much manual vetting. Now you can complete a loan application in three minutes and get approval in one second without the involvement of any humans.

Where does credit come from? Data. AliFinance is a product of big data, which is its foundation and core.

Centered on data and the internet, AliFinance's open microloan technology has become a solution for small and micro-enterprises. Online loans make information more transparent. AliFinance's bad loan rate is 1.02 percent, compared with China's banking industry average of 5.5 to 6 percent for small business.

Relying on data, rather than guarantees or collateral, is the main differentiating factor for AliFinance. This way of doing things not only makes it easier for smaller business to get loans, it also turns all the data they have accumulated on e-commerce platforms into a valuable credit check.

But this data didn't belong to any particular company. It was each person's own natural resource. Created by each user, it is a resource that would have been inaccessible before the online world. Today when you open your Alipay wallet, you can see your Sesame Credit Score, a kind of financial credit that provides much convenience in people's daily life. For example, if your credit score is over 700, you can use it to apply for a Singapore visa without needing to providing additional financial support documents. If it's over 750, you can apply for a Schengen visa without having to provide a bank statement. Data can mean your credit receives more respect because it better reflects your actual credit than a credit card.

Everything being online is a little like the private finance model of Wenzhou, a city in the east coast of Zhejiang province, where the majority of enterprises are small businesses.

Wenzhou's model also requires a different kind of data. But how is it gathered? The answer is very interesting, and it is very much localized. The lender hires an old lady for 10 RMB to sit outside a factory and count how many workers go in every day and how many trucks with shipments go out. From this data, the lender can evaluate whether the company is doing well or not, and then decide how much to lend to the factory owner. This is, in its rudimentary way, a form of data collecting and analysis. While it might be a bit unusual, it still counts as a data way. The only unfortunate thing is this all takes place offline.

In the past, credit scores were based on people's offline assets, i.e. houses, collateral, cars, etc. In other words, your credit was determined by your offline data. It is very easy to fake offline data. There are many ways to do this because everyone knows there are only a few types of things that are used for data collection.

The biggest challenge that modern banks face is processing all the offline data they have to evaluate your credit. This is obviously more difficult than hiring even a hundred old ladies to calculate. Banks' rating process is complicated and long, and the cost is high.

AliFinance's whole process, in contrast, is online, meaning everything speeds up.

If you have a goal when you are collecting data, it will be of limited usefulness because your goal will constrain your ability to imagine different ways to use the data.

AliFinance's accumulated data has no defined goal, it is just data that is collected through normal internet behavior. Therefore, this type of data allows a considerable amount of room for imagination.

But the change from having a goal to not having a goal is massive.

After finance went online, operating costs dropped dramatically, which allowed a service that only a small minority could afford suddenly became available to the masses. No longer do you have to be seeking 100,000 RMB loans for someone to pay attention to you. You can now get a loan for 1,000 or 2,000 or even just 1 RMB. As long as the unit cost of computing is low enough, microloans won't lose money, and the system can operate normally. At the same time, you need sufficient computing power so that you can calculate a result for a loan request quickly. If the online evaluation process took the requisite one or two months and could only meet with ten people a day as it does offline, online finance would certainly fail.

When you evaluate a new online product, it's important not to just look at its technology but also at what value it brings to society. Someone once wrote a thank-you letter to AliFinance for giving him a 1 RMB loan. This was something he never imagined. He might be the person who has received the smallest loan in the world. Regardless, that 1 RMB loan was not just a matter of borrowing money. It represented respect. In the past, this would have been impossible, but in the online society, this will happen over and over.

When AliFinance first launched, I said that it was creating a helicopter, not an airplane. In professional literature, no one calls helicopters heli-planes because they're not airplanes. The principle of airplane flight is very simple. Lift relies on two wings, which split airflow above and below them. The air also flows from the front end of the wings to the back end, so the air that passes over the top of the wing moves more quickly than the air that passes under the bottom of the wing. Airplanes compete on speed, altitude, and distance. These are the unique characteristics of airplanes. But helicopters use rotors, which rely on the main rotor and the angles of the rotors to generate lift and movement. The aerodynamics are totally different from airplanes. Helicopters are compared on which one can be stealthier and get closer to treetops. Airplanes and helicopters are two different varieties of aircraft. Air forces around the world normally do not have helicopters; they are under the jurisdiction of armies. Even though these two things seem the same, and both can fly in the atmosphere, they are totally different. By the same token, even though AliFinance and traditional banks look the same, they aren't. Not everything that flies is called an airplane.

Cities being online will also bring surprises.

Hangzhou is a very beautiful city. The September 5, 2016, G20 summit was held there, bringing attention from around the world. I live and work there. Cities are one of the amazing inventions of humanity, but almost every city in the world has traffic problems, and Hangzhou is no exception.

I have heard people talk about smart cities many times, just as I have heard about traffic jams. But I haven't heard an explanation for why smart cities haven't yet resolved the issue of traffic jams. With these questions, I took a look at the statistics.

Hangzhou had around nine thousand public buses with more than five hundred routes and seven thousand stops in 2016. Every day, the buses carry around 4 million people-trips. With this level of traffic demand, not planning would be impossible, as would simply relying on traditional methods to plan. If you don't have enough data, any type of planning will come to naught.

There are more than 3 million registered cars in Hangzhou, the highest per capita rate in the country. When these cars hit the roads, city planners have no idea where they come from and where they're going, so they're unable to optimize roads (space) and traffic signals (time) and other resources. Much of the time, you will find that the traffic cops' directions are different from the traffic lights. The reason is that automobile trip data hasn't yet been collected.

This problem exists in practically every city in China. Could we do something different from a traditional smart city? In April 2016, Hangzhou started the "City Brain" initiative to explore whether public data could help to tackle the challenges every city is facing. The initiative started as a nonprofit project with the support of local government and with participation of more than thirteen private companies.

It took the city and the private sector coming together using data as a resource to solve complex urban problems. I am proud to have been a part of this process. We will find a better way to understand cities when public data is being processed by City Brain.

For example, when people talk about traffic problems, the key data they're using is the number of registered cars in a city. Is this number the direct cause of all traffic problems? City Brain asked a different question. How many cars are on the street at a specific time of day? Surprisingly, with almost a decade of smart city development and billions of dollars invested around the world, no one had asked this question before. City Brain began by trying to answer this question. With the help of data modeling and computing power, Hangzhou City Brain gives us a surprising number: fewer than 10 percent of registered cars. Of 3 million cars in Hangzhou, about 300,000 are on the street during rush hour every day, and there are only about 200,000 outside rush-hour periods.

Solving a problem of 300,000 cars is very different from solving a problem of 3 million cars in terms of the physical resources required. We might not need to use as much city land to solve traffic jams if traffic can be coordinated more efficiently. In this instance, data resources play as important a role as traditional resources, such as land.

In Hangzhou, City Brain is coordinating traffic signals with calculated road traffic to provide digital driving guidance for drivers. This helps to solve much of the problem of traffic jams in the most populated areas, such as the business, hospital, and school zones. Today, finding a parking space is much easier than before, with the real-time parking space information provided on digital traffic boards in the city. With the help of the cloud, data is driving better traffic management over the allocation of additional land space for transportation. When data is processed through the brain in this way, it will become a source of prosperity for the city.

Through the construction of the City Brain, the city of Hangzhou will be exploring cloud computing and big data solutions for cities in China and around the world. In December 2017, I gave a keynote address at the Annual National Planning Conference and envisioned how City Brain could help every other city save 90 percent of the resources they were currently using. That would mean we would need only 10 percent of water supplies, 10 percent of the land usage, and 10 percent of the electricity used now to survive in our city. It is a dream of the sustainable development of cities.

But it isn't just a dream. It is an unprecedented exploration. This is what a sustainable city truly means. One thing about the future that fills people with excitement is how data will permeate every aspect of our lives, from education to health care. The small step of City Brain is not only a giant leap for long-term city development, it is a milestone for the value of data in society.

Chapter 7

From Menlo Park to Yunqi Town

In August 2014, a friend sent me an essay posted on WeChat by Guohui Zhou. I was very moved by the piece and was surprised to find that what I had been doing had such an impact on the things around me.

After living in Hangzhou for such a long time, the city's heavenly beauty practically had become mundane for me. It was to the point that I needed to learn to be better at appreciating subtlety or change my perspective, such as by viewing the city from an airplane, where West Lake looks like a pearl embedded in the verdant earth and the changing seasons engross and inspire contemplation.

Hangzhou's beauty is never without change. The development of information network technology is giving birth to new industries and a new city. That is the beauty of flow.

The people of Hangzhou all know the place called Yunqi, a scenic site not far from Jiuxi, known as Yunqi Bamboo Path. I remember more than ten years ago in fall, going uphill from Meiling Tunnel, looking at the

Longjing tea plantations surrounded by green mountains and sparkling waters. Pavilions perched among mountain homes in a heavenly vista. Going down from Yunqi, the bamboo path was alive, a cool breeze was gently blowing, and I was filled with contentment and joy. If I remember correctly, the famous Dharma master Hong Yi became a monk in a temple nearby.

But what I was unaware of was that a new place called Yunqi Town had appeared near West Lake. It is located less than ten kilometers from Yunqi Bamboo Path (which used to be one of ten best scenic spots on West Lake), not far from the China Academy of Art Xiangshan Campus and the Zhejiang Academy of Music, still under construction. Last Thursday, I went to Yunqi Town on a field visit with the governor of the province, and it far exceeded my expectations, reminding me of the saying, "What you learn from books is never as great as what you learn from experience."

Yunqi Town is neither Yunqi in toponymy nor an actual town on administrative maps. It was a no-name place and is a cloud computing campus. With the annual Aliyun Developers Conference, the secluded mountain valley was hosting a "green revolution" that might impress China and the world. The event brought together industry leaders, tech experts, corporate executives, and internet developers to discuss the technology and applications at the intersection of cloud computing and the internet.

Starting in 2010, the developers conference has been held every fall. This all helps promote the development of China's cloud computing industry. Last year, when the tangerines were ripening, Yunqi Town hosted more than five thousand internet, cloud computing, and big

data experts from around the world. I believe that day
the whole valley must have been full of hubbub. I can
visualize the scene before me like in a movie where a
single spark starts a prairie fire.

The lead planner and builder of Yunqi Town is
Alibaba CTO Jian Wang. He told the governor that
many people had doubted his planning, asking, "Who
would come to this no-name place for a conference?"
But this conference is not like others. Participants
come because they love innovation and want to start
start-ups, not just because there is a conference going
on. In 2013, at the Aliyun Developer Conference, Jian
Wang said Aliyun is possible because of the intense
deep devotion of so many developers. But beyond
knowledge of technology, what everyone expects is
the true friendship and openness that allow people to
do great things together in the future.

Even though it was already growing dark, after the
governor listened to the reports he went in the rain
to the Apsara 5K Base, which was running an Apsara
cluster with more than five thousand computers. With
home-grown technologies, the Apsara initiative began
on October 18, 2008, and was deployed for public
operation on August 15, 2013, making Aliyun the first
Chinese company to develop its own large-scale general computing platform. This was also the first time
a commercial company provided 5K data processing
services to the public.

A stone statue was erected near a pristine river. On
it is written: ON OCTOBER 24, 2008, A GROUP OF IDE-
ALISTIC BROTHERS SPENT 1,757 DAYS AND NIGHTS AND
UNTOLD BLOOD, SWEAT, AND TEARS TO WRITE CODE
SO THAT THE DREAM OF CLOUD COMPUTING COULD

BE REALIZED. Below the pillar is engraved Jian Wang's signature and all the names of Apsara developers. Jian Wang said this is not just a dream and a promise. It is something we need to turn into a reality through our own actions. In fact, dreams, promises, and actions are the three cornerstones of all achievements.

When I travel abroad, my dream destinations are not big cities like Tokyo, Paris, and New York. I am drawn to the Nile, the Amazon, and the Alps, the natural, ecological, and cultural small towns. Hearing from the governor, Zhejiang was making a plan to build a hundred unique towns for innovations just like what Yunqi was doing. That made me very excited.

Even though I thought up the name Yunqi Town and had personally spent several years working hard there, this article was still inspiring and made me think of a park on the other side of the world.

A Park Lights Up the World

On the map, Menlo Park is translated as Menluo Pake in Chinese. It is one of the most educated cities in California and all of the United States, with around 70 percent of the population having a bachelor's degree and up. But in New Jersey, there is another Menlo Park. Even though this is less well-known in China now, an invention there lit up the whole world.

When people think of electricity, they immediately think of Thomas Edison, mainly because he invented the light bulb. Edison was an inventor and an entrepreneur. All of his start-up capital came from J. P. Morgan. But the most amazing thing Edison did was that he helped to introduce electricity into cities. Not only did he invent the incandescent light bulb, he also developed generators and large-scale electricity transmission. Edison was the first to turn electricity into a public utility, and when he electrified Wall Street, electricity quickly became a form of infrastructure people couldn't live without.

Electricity brought about even more inventions, like the washing machine, hair dryer, and refrigerator. This is the effect public utilities and infrastructure have on innovation. Computing is the same. As soon as computing becomes a public utility, there will be a large space for innovation.

Edison made real contributions to humanity, like the light bulb and public electricity, but he also made big mistakes along the way. One example was provided by the "war of the currents" as described in detail in the novel *The Last Days of Night* by Graham Moore—when Edison, seeking to commercialize direct current electricity, which he had developed, went to great lengths to sabotage his former employee and rival Nikola Tesla, who had developed alternating current as a more efficient and economically viable alternative. From this perspective, it is possible to conceive of the internet today as a future hindrance to development.

Edison's company was the Facebook or Google of its day. It was the most advanced and productive, and young people like Tesla from all over wanted to work there. The East Coast of the United States was a hub of innovation at the time, and East Coast innovation went far beyond this. When, decades earlier, the United States built railroads across the country, it wasn't because entrepreneurs already knew exactly the long-term impacts of such technology. They did it because they were thinking about making travel and the shipment of goods across the country more convenient. Standard Oil founder John Rockefeller understood that one of most important commercial values of the railroads would be to transport oil, and he negotiated with the railway companies to control the main lines at very low cost. The railroad companies discovered later that buying and selling oil was far more profitable than operating railways. Similarly, when we first built the infrastructure for the internet, email was the only use. Later came Web surfing and e-commerce. While fiber-optic cables are expensive to lay but don't make much money, internet companies can easily turn a profit.

But what was so amazing about Rockefeller was his innovation. At the time, refineries were places full of pipelines and were considered high-tech. When major railway companies announced jointly that they were raising prices, Rockefeller wondered why pipelines couldn't be used to transport oil. Now oil pipelines are standard in the oil industry.

Events never occur in isolation. In electricity, Edison's boon was Rockefeller's bane, because people no longer needed oil lamps. Fortunately, Ford's large-scale production of the internal combustion engine became a driver of industrial development, and Rockefeller was back again on top. World development always surprises with magic like that.

At the turn of the twentieth century, innovation on the East Coast of the United States defined a generation. The internet is now playing that role today, and a transformation is brewing that will be more exciting than that sparked by Menlo Park and even more expansive than the one triggered by Silicon Valley.

Yunqi, Built for Developers and Start-ups

In the same way Menlo Park was just a small town outside New York City, Yunqi was a small cloud computing industrial park outside Hangzhou. I named it Yunqi Town. In order to be closer to entrepreneurs who believed in cloud computing, Aliyun formed an irrevocable bond with Yunqi. When 2013 was almost upon us, Jiang Liu wrote down his observations about Aliyun's 2012.

At the end of 2011, I went to Aliyun's headquarters in Hangzhou to speak with Jian Wang and members of his core team. This was the first time I had taken an up-close look at Aliyun. At the time, they were working on both a cloud computing platform and a mobile operating system. While it was grand in scope, the achievements were still not yet visible and there were serious doubts from inside and outside Alibaba. After that visit, I wrote an article entitled "Pursing the Dream of Clouds," in which I wrote: "Aliyun's strategic direction is correct, and the boldness to pursue indigenous innovation is worthy of respect."

But at the time, I also observed that the difficulties Aliyun faced in terms of both technology and business were not small. Jian Wang and his team were under immense pressure, and while the future was bright, the road was winding. In reality, by the time the first line of code was written for the Apsara platform in October 2008, they were on a road of no return, as Wang said. After years of reshuffling personnel, fewer than ten of the original team members remained.

In 2012, tech giants like Google, Apple, Facebook, and Amazon were in an all-out war over everything from hardware to software, mobile to cloud platforms, and developers to digital media content publishing. Chinese internet and IT companies were also fighting openly and covertly over search, cloud computing, open platforms, e-commerce, and payment. It wasn't surprising that, for a platform enterprise like Alibaba, competition in various verticals was unavoidable.

In 2012, it became common in China for internet companies to start making cell phones. The focus of media coverage regarding Aliyun was mainly on its mobile OS, YunOS. Its collaboration with and then separation from phone maker K-Touch was followed by the cooperation with Hai'er, then in September its cooperation with Acer was scuttled under pressure from Google. In a short span of time, YunOS became a business unit independent from Aliyun, and it found more partners to work with.

But outside the media spotlight, Aliyun's cloud computing business was doing well. It had become the domestic industry leader in both size and product lines. On November 9, 2012, at the Aliyun Developers Conference, I spoke again with Jian Wang. He revealed that tens of thousands of websites, apps, and games had been built on Aliyun, supporting more than 100 million user hits.

Aliyun's lead in some respects reflects the awkward state of China's cloud computing. We know that AWS is the most successful global cloud computing service provider. It has developed a thriving ecosystem from its Infrastructure-as-a-Service to its Platform-as-a-Service offerings. Amazon has provided the basis for the rise of many of the most successful recent mobile and internet start-ups in America.

Meanwhile, cloud computing in China looks hot superficially. The Chinese Electronics Institute and CSDN co-hosted the 4th China Cloud Computing Conference, which considered the year of 2012 as the beginning of the Chinese cloud computing era. But aside from the government-backed "data center building fever," there are almost no companies, aside from Aliyun, that are determinedly investing in creating a general purpose cloud computing service as a public utility.

Jian Wang said that when Aliyun was being established, he originally thought about naming it General Computing Corporation. Like General Electric for electricity, Aliyun from day one has been focused on general computing. Most companies underestimate the difficulties surrounding the technology of general computing and the business model of a competitive service. The barrier to entry will be lowered significantly by working in some selected verticals, but it would be unlikely that a vertical could transition into a large-scale general cloud computing service platform. No software stack exists for such a computing service platform. This is also the reason Aliyun has taken on considerable risk to go around existing open-source software and write its own Apsara system from scratch, thereby bringing together Amazon's and Google's advantages. The number of companies today that can provide comprehensive cloud computing services isn't increasing; actually, it's shrinking. User diversity has raised the barrier to entry for general cloud computing services.

Jian Wang explains his company's differentiating factors this way: "Aliyun's only business is cloud computing. Cloud computing has two forms, one is integrated into other services, while the other is not. Aliyun is not integrated into any other kind of business; it's only cloud computing. If cloud computing is bundled with any existing business, it will definitely become a secondary product, whereas users demand innovation with new computing power. Aliyun luckily is not integrated into any existing Alibaba business, and users can develop whatever kind of service they want.

"Aliyun is a truly open platform. What do I mean by 'open'? I have thought about this question for a very long time. Being open means allowing small firms to be able to do what only big companies were once able to do, and vice versa. This is the direction in which we are moving."

Why does Aliyun differ from other platforms in this way? Wang believes it has to do with the company's DNA. The DNA that stands up and fight for what is right, for "small and beautiful" individuals, small- and medium-sized enterprises and other comparatively weak entities. Using computing power unleashes their creativity and creates new types of businesses and helps them grow.

At Aliyun Developers Conference in 2012, the presentation of Wang's speech had only one slide, entitled "Developers—Apsara—YunOS—Developers." The presentation started: "There shouldn't be a difference between leaders and guests. At the developers conference, there is only one person—the developer." When asked why Aliyun had invested so much in the developer community, hosted a developer conference, organized programming competitions on cloud, and promoted a developer certification plan, Wang said: "Because Aliyun was born for developers."

"Developer" at Aliyun really means the developer of computing power. While the YunOS operating system was created to bring mobile hardware and software developers together, Aliyun's mission is to provide technological and computing resource support to allow an individual to be able to do what only a large company could do in the past. Meanwhile, cloud computing will force traditional companies' work processes and organizational structures to change by helping them reach the level of efficiency and flexibility of the internet and SMEs [small- and medium-sized enterprises].

"If there were no Ford assembly line, electricity would have had no value." Wang likes to compare the changes brought about by cloud computing with those of electricity. Electricity changed many industries and the whole face of society. Aliyun is accustomed to evaluate a business through its impact on and contribution to society. Profit is the result of this impact.

2012: A difficult year

Outsiders might not know that 2012 was not an easy year for Aliyun.

This was the epitome of China's cloud computing in practice. Aliyun's first cloud computing service launch once made top developers at small and medium-sized firms very excited. In the past, their choices were either Chinese IDC servers, which were difficult to run and maintain, or cloud services outside of China, which meant access speeds were incredibly low.

But after they got onto the cloud, early users discovered that even though Aliyun's network quality was outstanding, the service was unstable and there were frequent malfunctions. Poor input/output performance was even more difficult on users. After these problems were resolved as the platform improved, Wang discovered an even bigger problem: customer service.

Aliyun had put all of its energy into tech R&D, making software to run the platform. With less service awareness and a lack of service experience, many quality-of-service issues arose as a result after Aliyun began providing services.

Once, Wang heard that a small client, TeamCola, which did team collaboration with SaaS, decided to leave the cloud because of poor service. When Wang asked his staff about this, no one seemed to know that it had even happened. It turned out Aliyun's staff cut services to one of TeamCola's cloud servers without notifying them in advance, simply because the client hadn't completely registered a domain name to comply with local regulations. "Even the most bureaucratic, traditional IDC would have at least given us a phone call before terminating service," TeamCola founder Xueliang Shen later recalled. "This type of service completely destroyed our confidence." Wang was furious. He carried out an office shake-up and told his staff: "If we become an arrogant company, we won't be far from the end as a company."

Aliyun made both product and organizational changes. With more integration of the platform components with service products, users are able to problem-solve on their own. By splitting up the responsibility for running the platform, the developers could focus on their products and let the maintenance staff take care of upkeep. The two groups have a mutually reinforcing relationship that improves service quality. Additionally, Wang increased the number of product managers and exhorted them to talk directly with their clients.

This reshuffling improved Aliyun's service significantly. From my observations, Aliyun's Sina Weibo accounts, particularly their customer satisfaction center account, are the fastest and most efficient at responding to customer complaints among all cloud computing service companies. Many of Aliyun's employees are very active and full of energy, and almost every Weibo post mentioning Aliyun is answered by an Aliyun representative.

Clients reacted positively to this improvement. In August 2012, TeamCola was back with Alibaba Cloud. Xueliang Shen said: "The facts show this is the right decision. So far, Aliyun's service has been operating at 100 percent service continuity. Further, we will move all of our services to Aliyun, which is essentially placing our company's future in the hands of the Aliyun platform." I interviewed another user who had many complaints about Aliyun, but he was quick to point out that their customer service staff had a good attitude.

Further, Aliyun didn't use the traditional IDC sales model through salesman offline, instead relying only on online self-service on the Aliyun.com website. Business still grew rapidly. Wang believes that having this model was one of the great accomplishments of 2012. It is more effective to have the tech staff promote products to users and handle and resolve technological issues, rather than having a marketing team that doesn't completely understand what they're selling.

"Clients taught us many things, and even changed my worldview," Wang said, adding that the business growth and increase in clients in 2012 greatly benefited Aliyun. "It is the developers and clients who are really teaching us how to do cloud computing." This is also one of the reasons Wang puts such an emphasis on the developers conference and building a community of developers.

Controversy from within and challenges from without

People who have met Jian Wang can tell he is a very prescient and strategic thinker. A clear example of this would be that in 2009 when Aliyun was founded, their mission statement was: "To make an advanced and data-centric cloud computing platform." This was long before big data had become a hot topic in China.

Within Alibaba, few fully understand Wang. On Zhihu (a popular tech forum in China), in a thread entitled, "What kind of person is Jian Wang," more than one Alibaba employee expressed the opinion that they often did not understand what Wang said. From my interviews, I also learned that aside from Jack Ma and former strategy studies professor and Alibaba chief adviser Ming Zeng, many executives in the company don't always follow what Wang says and have to ask Zeng for an explanation.

One of the reasons people had doubts about Aliyun was that no division within Alibaba Group actually used Aliyun computing services. This meant there was no compelling internal case study of the application of this type of business. Inside Alibaba Group, Aliyun and other businesses' R&D teams were often at loggerheads about technology. The reasons are not hard to understand.

Moving vertically optimized and stand-alone core business systems onto Aliyun, a relatively new and not easily compatible platform, was a major challenge and clashed both in terms of the technology itself and the attitudes and interests of the personnel. The evaluations of Jian Wang and Aliyun were tainted by this clash, and for a time there was fierce debate and even blunt criticism of Wang within Alibaba.

In August 2012, Wang was appointed as Alibaba Group CTO while remaining president of Aliyun as well, meaning that in addition to cloud computing, he was responsible for planning, setting, and implementing the group's technology strategy. This promotion allowed Aliyun to play a much more important role in the future of the group's technology R&D direction. It is also worth paying attention to whether the technology of Alibaba Group's various business lines will integrate well with Aliyun. If there are serious errors and malfunctions while one of Alibaba's core businesses migrates to the cloud, it would be a major setback for Aliyun.

Aliyun faces many clear challenges from outside as well. China's particularly regulated environment and internet restrictions are challenges. And as a path breaker in an emerging industry, they need to address problems that no one else has before.

From a macro perspective, it isn't particularly clear how a cloud computing business could make a profit in China's unique environment. While Amazon's cloud computing platform appears to be on fire after over a decade of rapid growth, the division's income was less than $2 billion in 2012, and it had low profits. In comparison, iTunes and iCloud, which use considerably fewer resources, have much better profits and income. The draw of this type of opportunity cost could be a hidden but more serious challenge to Aliyun's ability to continue on its current course, increase investment, and improve products and services.

AWS is without a doubt the key player in the global cloud computing industry. As far as China is concerned, the success or failure of Aliyun will have a considerable impact in same way. Let us wish the best for Aliyun.

In this environment, Aliyun has taken on an abiding responsibility to innovators and entrepreneurs. This is the reason I wanted to have the name Yunqi. In Chinese, Yunqi means "the place where the cloud dwells."

As Jiang Liu noted, the only slide for my speech in 2012 was "Developers—Apsara—YunOS—Developers." Later, I stopped using any slides for my speeches. Now I speak from my heart.

The Apsara Conference: A Computing Conference

Hangzhou is a city that puts China in the cloud.

In January 2013, we worked with Huatong (an IDC service company) to make the first data center in Hangzhou's Zhuantang Cloud Computing Park. We later named this park Yunqi Town. We came to an agreement with the park's management committee and built up the number one base camp for innovation and entrepreneurship. From then on, Yunqi Town has had special meaning to me.

I decided to move the 2013 Aliyun Developers Conference from a hotel in Hangzhou to Yunqi's open-air field, a place without any conference facility. Nearly five thousand people attended the tech conference, which made me grateful for Hangzhou's wonderful weather and all the developers who came. Apsara's first cluster with five thousand servers was deployed at Yunqi's data center, lending a unique subtext to the conference. The Yunqi Alliance was also formed at this time. The following is something a CSDN reporter once wrote:

In 2013, the cloud computing industry was estimated to be worth 110 billion RMB. In 2015, predictions have that number soaring to 750 billion or 1 trillion. Innovative applications based on cloud computing have penetrated into many sectors. Finance, traffic, electricity, manufacturing, and telecom operators among others have fully embraced the internet to seek technological breakthroughs, while online gaming, mobile internet, the internet of things, and smart cities are thriving.

From October 24 to 25, China's leading cloud computing firm, Aliyun, hosted the 2013 Aliyun Developers Conference at their innovation and start-up base camp. The conference theme was "The Butterfly Effect of Cloud Computing." The conference gathered over 5,000 Chinese and foreign experts in cloud computing. There were five sessions covering mobile internet apps, gaming, finance, developer services, and YunOS that gave developers and companies in the cloud computing space a chance to exchange ideas about technology and innovation. Beyond that, the conference also gives industry observers an opportunity to understand the newest developments in cloud computing and its applications.

After the final round of the developers' competition, thirty representative companies (including Aliyun) announced the launch of Yunqi Alliance [the cloud computing ecosystem alliance). At the first meeting of the founding members, Jian Wang said: "From day one, Aliyun has always relied on its friends for support. We hope the Alliance will be able to work with more companies on a deeper level to promote changes in the industry. Entrepreneurship and innovation based on cloud computing allow every developer to stand at the same starting line as big companies, gives them the

same capacity as the major firms, and lets them innovate the way they want. At the same time, the cloud
allows traditional enterprises to have the same capacity
for innovation as internet companies."

If you have the opportunity to visit Yunqi, you
can see the names of these enterprises on stones lining
the roadside. I am grateful to all of these companies for
their contributions to the development of cloud computing. Without their efforts, cloud computing would
not be the way it is today.

Going back to May 2013, CSDN published yet another
status report on cloud computing:

Cloud computing is developing very quickly, and
more technological and non-technological problems
are arising. The "biggest victim" of these issues are the
cloud services users. Facing criticism from all sides,
service providers have struggled to respond. AWS is
far away and Google is behind the Great Firewall, but
the Playcrab vs. Aliyun incident we can examine.

In the past two weeks, China's cloud computing
industry has been buzzing about one event: Kai Ye,
CEO of Playcrab, the maker of Dazhangmen [a well-
known video game], launched a Weibo rant complaining about technical and service issues for the company's
game on April 27. Many people came to his defense
online, and the cnblogs.com posts complaining about
Aliyun came to the fore. There were even some solutions to help companies to migrate to other cloud
platforms from Aliyun. Aliyun's initial response was
a request to speak privately with Kai Ye, but they did
not receive a satisfactory response and later maintained

silence. It was not until Jian Wang personally flew to Beijing to talk with Kai Ye on May 6 that the company announced the two parties would continue to deepen their cooperation. This seems like a minor issue, but it had an important effect on the development of the Chinese cloud computing industry. A situation like this would seem a common occurrence in the United States, but this time it was happening in China, among all of us.

As American cloud computing applications broaden, start-ups' understanding and use of cloud computing platforms has reached a new level; more problems have arisen at the same time. For example, in February, James Somers of the website Rap Genius reported that the cloud computing platform Heroku had seen a user information leak after redesigning its routing system and that a class-action lawsuit was in the works. Then came the March report that probed AWS's "copycat" partners like Newvem, Zencoder, Rightscale, Cloudyn, and Cloudability. But in China's comparatively quiet cloud computing industry, a four-day-long server shutdown and data loss from Shanda [an early cloud service provider in China, which suspended its service in 2013] did not spark much controversy. However, when Aliyun launched the "My Stories with Aliyun" theme on Weibo, which mentioned Playcrab's Kai Ye, the complicated details began to surface.

The response came from Kai Ye on April 27 at 11:04 on Weibo. "We used over twenty servers on Aliyun. In half a year, there was one time when all of our servers went off line, two times when multiple harddrives suddenly became read-only, three times when the hard-drive input/output suddenly became very slow (the explanation they gave us was another virtual server on the same physical server was taking resources from us), and one time customer service restarted our servers without letting us know. There were more than ten times when our systems would not respond, and our phone calls were never answered within five minutes. We have already prepared to migrate all of our systems off of their servers."

Kai Ye's response was brimming with anger. Later, many other cloud computing users brought up similar concerns. "They kept making us change our systems in order to adapt to their problems. But we operated normally using different clouds and IDC. We are always optimizing our systems, which mainly ran well on other cloud platforms and IDC machines. But they never thought about their problems much less resolved them. They would only shirk responsibility, meaning their products and services did not improve, but worst of all was that their customer service representatives had a bad attitude. Further, there were problems too frequently, and they had staffing issues after they bought Wanwang [an instant messaging tool for customer support]. The newcomers didn't know who they were working with or what module they were responsible for. I feel like they really weren't concerned about growing this business."

While Kai Ye lodged all of his complaints at once, the blog posting on and by cnblogs.com was far more detailed and consistent.

Details from cnblogs

Cnblogs.com is the largest online developer community in China. Cnblogs.com represents a type of cloud

computing user and website with considerable potential. And every problem the company encountered
after migrating to the cloud was recorded on a daily
blog diary by the site's CEO, Yong Du. Many discussions were of value to other cloud users.

But in fact, just as cnblogs readers have said, from
the time they were preparing to migrate to Aliyun
at the end of last year to now, there have been many
problems. According to Yong Du, technical issues
have been the main problem, while the service has
been deserving of praise.

CSDN: Reading your series of articles, I have been
moved. What do you feel has been the most difficult
part?

Du: The cloud servers' disks' I/O is unstable. When
we started using cloud servers to run our database, it
worked well if the disks' I/O was stable. But later there
were several server problems with our database because
of the disk I/O. And our search product remains trou-
bled by the cloud server disk I/O problem.

CSDN: Why have there been so many problems?

Du: There isn't enough consideration given to the
situation of users, like how to properly resolve the competition for computing resources by different users.
And there isn't enough emphasis placed on system
stability.

CSDN: Have you encountered bad service (no
response, unanswered phone calls, etc.)? What do you
think about Aliyun's service?

Du: We haven't encountered any problem with
not getting a response. The most praise-worthy aspect
of Aliyun has been their service. Through our interactions with customer service, we get the feeling that

they are actually trying to solve problems but not waste time.

CSDN: From migrating to the cloud to using the cloud computing platform, what has been the most standout experience you at cnblogs have had with Aliyun?

Du: It is amazing that we have the freedom to expand our computing capacity at any time. We can increase or reduce the number of servers we use whenever we want.

CSDN: If you had the opportunity to meet with Jian Wang, what would you ask him? What recommendations would you have for Aliyun?

Du: I would ask how good their virtualization technology is compared to other companies in the industry. My number one piece of advice would be to make stability the top priority, because the cloud supports tens of thousands of websites serving hundreds of thousands of customers. Also, the Aliyun website user experience is really very bad.

CSDN: AWS and Heroku also often have issues. What is the difference between Chinese and foreign companies? What do you think of the whole cloud ecosystem? What are recommendations for building a domestic cloud ecosystem?

Du: In terms of having problems that affect users, there's no difference between China and overseas, which shows that the technology is not yet mature. Why is it that when one power plant has problems and cannot produce electricity, we can still use electricity? My understanding of the cloud ecosystem is lacking, but I think the cloud service providers play the key role. Once they start focusing on trying to become like power and water utility providers that don't have any supply problems, everyone will start using the cloud.

CSDN: Have people recommended you migrate off Aliyun? Do you have any plans to do so?

Du: We can make evaluations only based on our own personal experience. We haven't left Aliyun because we believe they can resolve our issues. We don't want to leave. I believe any company that leaves a cloud platform isn't doing so because they'd planned it, but because they were forced to make that decision. And cloud computing users will all tell you that moving off is very difficult.

Cloud: to be or not to be

Is moving off about moving to another cloud or returning to IDC?

As cloud computing services have become more widespread, this question has arisen more and more. But regardless of whether you're talking about China or other countries, there seems to be only one solution: migrate. Even though many companies are currently adopting mixed cloud, utilizing their own servers and the cloud, as soon as they experience the benefits of the flexibility and speed of the cloud, they are reluctant to go back.

In reality, the stability and service quality of cloud computing platforms have long been controversial, both in China and the United States. Recently there have been a spate of articles on this subject on CSDN:

"June 14, 2012, Amazon Cloud Services analysis"
"The story behind the cloud server breakdowns at Amazon, Google and Apple"

"Crisis hits AWS data centers. What is the public opinion?"

"AWS's eastern region breakdown hits Netflix and other websites"

"Impact lasts 6 seconds, Heroku's routing redesign is revealed and the costs are high"

But at the same time, there are some users who have developed and open-sourced very useful technology and applications. One type is Netflix and Instagram, which run on AWS. Not only are they cloud users, they are an important part of the ecosystem, and even make up for some of the shortcomings of AWS. Netflix's open-source Janitor Monkey (a cloud computing management tool) and Eureka (a load balancing and failover tool) have already helped countless enterprises take better advantage of AWS.

Looking back on that period today, the costs we paid to build up this industry were worth it, and I'm very fortunate to have been involved in these events. At Aliyun's first Developers Conference in 2009, there were more than four hundred attendees who were on the cutting edge of China's cloud computing industry. By 2013, more than five thousand people came to the conference, evidence of the development in the industry.

In 2014, I had a colleague print out Yong Du's blog posts and turn them into a pamphlet we called "The Progress Collection." I told our staff that this was Aliyun's bible. In 2015, Yong Du and cnblogs.com moved into Yunqi Town. It is a team of ten people, but running an online community with one million developers.

It was in October 2015 that Aliyun's Developer Conference officially took the name of Apsara Conference, and hosted twenty-two thousand people, 80 percent of whom were under thirty-five years old, with the youngest person only sixteen. These twenty-two thousand people represented all of China's thirty-two provinces and administrative regions, including remote Tibet and Xinjiang. Of course, there were also many foreign friends as well.

In 2015, people in Hangzhou and other major cities around the country could see the advertising slogan at the airports: "Computing for Incalculable Value." Starting from Yunqi, Hangzhou is the city that will put China on the cloud. I believe that in the near future, Hangzhou will become a city that uses computing to change the whole world.

Yunqi's Journey with Computing

Zhongguancun Village isn't a village. Yunqi Town isn't a town.

Hosting marathons has become a new trend for Chinese cities, and it is something that seems to have some link with the phenomenal rise of entrepreneurship. In November 2015, the Zhoushan Islands (famous for Mt. Putuo, one of China's four sacred mountains) held their first marathon, which attracted thousands of technologists around the country. Hosting the marathon was my idea. In a visit to the islands, I discovered that the PM 2.5 pollution index there was only around a 10 on average, while Beijing was a few hundred at that time. "Why not have a marathon here?" I thought. "Why do we have to have marathons in a place where the PM 2.5 index is at 200 every day?"

TMTpost, an IT media outlet in Beijing, interviewed me before the opening ceremony of the marathon.

Jian Wang organized an event "Running into Zhoushan Islands from Yunqi" and crafted a slogan: "Yunqi:

If you're not serious, please don't bother. Zhoushan Islands: if you're serious, see you there." Yunqi is a cloud ecosystem-focused industrial park for start-ups and innovators. It has already attracted many internet entrepreneurs. What Wang means by his slogan is that just as people who love running know where to go to participate in marathons, so too should serious entrepreneurs find places that allow them to unleash their creativity and imaginations.

Wang believes that Beijing has already become a place where people just can't do serious start-ups, meaning there are now a lot of entrepreneurs there who aren't serious. The cloud computing industry's development is built on providing lower-cost technology to internet start-ups and releasing entrepreneurs from the constraints of geography. The entrepreneur class will form a different gathering place, which is to say that there will be an economic region more suitable to start-ups and innovation other than Beijing, Shanghai, Guangzhou, and Shenzhen. Last month, 20,000 people attended the Yunqi conference, and now they are bringing their energy to Zhoushan. Wang hopes that everyone can stop focusing exclusively on Beijing, Shanghai, and Guangzhou when they look into innovation and entrepreneurship.

Where China's garages are

We often say that the air in Beijing is bad for running a marathon, but no one says where else they should host one. What am I trying to say? We always talk about the start-up ecosystem in Beijing, Shanghai, and Guangzhou. The ecosystem is important indeed. It means the pressure put on people and the constraints placed on their imaginations. This is not a matter of air-quality standards, but rather of the outlook an individual may be able to have. I once talked for a long time with a person who was doing a start-up of office sharing like WeWork, and he said the main innovation and competence of his company is that the offices they rent out are all with standards for PM 2.5 air quality. If that is our biggest innovation, then it's not really innovative at all.

Is innovating in a garage the same as in an office building? I have seen many early-stage start-ups in Beijing that move from one office building to another. These entrepreneurs don't have the sense of ownership that people have when they start doing something in their garages. That is why I said that the Zhoushan Islands could establish a place for geeks to feel at home.

The internet spirit is to share and spread, not to centralize. We all mix up physical things with the virtual. The result of being virtual is that distance and location no longer matter, but the sense of a physical existence is still important, and certainly could be found in many other places than Beijing.

The turning point for the cloud computing industry has arrived

In the past, when foreign media would describe China's development, they would say something like this: "China is using one-third of the world's steel, half of its water. . . ." But in ten or twenty years, foreign media will look at China again and talk about how much of the world's computing power we're consuming. This would be a true change.

Looking back on the evolution of America's economy, you'll discover that most of the biggest companies developed and grew in Pennsylvania, then the center of growth moved to New York. After the invention of the telegraph and radio, it began shifting west, and finally Silicon Valley emerged. This westward move happened without anyone actively coordinating it.

As cloud computing spreads, the current capital of the American cloud computing industry is Seattle, where Microsoft and AWS are headquartered. Aliyun has set up an R&D facility there too.

This move has been a natural occurrence, not something out of a government plan. Economic and innovation centers shift without anyone willing it.

If you look at our current infrastructure, you can see that you don't have to be in Beijing, Shanghai, or Guangzhou to run an internet business. The greatest difficulty tech companies face is finding talent; geography is not important. People can gather in two ways: either in online communities or in real life. In the next twenty years, there will be a new place in China that will serve the role that has been played by Beijing and Shenzhen over the last thirty years.

Although there are no longer geographical constraints on innovation and entrepreneurship—it doesn't matter where a new product is created—a physical sense of being is still an important thing, and entrepreneurs require a memorable place to belong to. This marathon is an effort by the Yunqi Alliance to bring people who may make high-value innovations in the future closer to the Zhoushan Islands.

This article was later given the headline, "Looking at Yunqi and Zhoushan Marathon, China's Economic Center Is

Shifting." Jiang Liu also recorded what he had observed in a similar article posted at about the same time.

It is without question that Amazon's cloud computing services have played an extremely important part in the development of the whole industry, and in China that role is being performed by Aliyun. In 2013 and 2014, Aliyun has been at the forefront of the discussions in the domestic cloud computing industry.

In 2011 and 2012, I interviewed Jian Wang twice and published two articles recording the successes and failures Aliyun and Wang himself have experienced over the different stages of the company's development. In the second article, I concluded by saying, "It is without question that Amazon's cloud computing services play an extremely important part in the development of the whole industry, and in China that role is performed by Aliyun."

What I didn't expect was that, a year after these articles, the global cloud computing industry would undergo a major transformation.

The story that attracted the most attention was that Amazon had received a $600 million contract from the CIA, meaning that they had strongly entered a key market space—government. More dramatically, IBM sued the CIA over this, and was still unable to win back the government's business.

Furthermore, Microsoft and Google abandoned the strategy of only focusing on PaaS and later made their presence in the IaaS market, triggering a chain reaction. On June 6, 2012, Microsoft announced its first IaaS [Infrastructure as a Service], though the company was still using mixed cloud to cover. Meanwhile, the same month, Google revealed its IaaS platform GCE at the I/O Conference, taking clear aim at AWS. Yet it would be almost a year later that both companies would formally launch the services of IaaS in April and May 2013. They were five years behind Amazon. In August last year, according to figures from Gartner, AWS's computing capacity was five times greater than that of its next fourteen competitors combined.

But the two tech giants certainly have resources of their own (technology, of course, but also hundreds of millions of dollars). As long as they make a decision at the top to do something (a decision has to be made by the CEO or number one in a company), they'll be able to catch up very quickly, regardless of the market or the product. In January 2014, Urs Hölzle, Google's VP for infrastructure, sent out a shocking memo saying his team would stop focusing on internal "clients" like Search and Gmail and would instead put their energy into reaching out to new clients by building a public cloud platform. In February 2014, Satya Nadella moved from his role as head of cloud computing to CEO, and he made cloud computing one of two of the company's core strategies. Azure, Microsoft's cloud computing OS, immediately got a boost both in terms of product development and marketing.

The biggest competitive advantage big companies have is their large amount of fixed assets (they invest billions of dollars every year), which allows them to engage in fierce price wars. In March 2014, Google slashed prices of its services by 32–85 percent, and Amazon followed the next day, with Microsoft following suit only a few days later. The latter two, however, didn't lower prices as much as Google. The result of price wars is that "when big companies are fighting each other, small companies get hurt." Rackspace, an independent cloud provider, was the first to fall: they refused to cut prices, and amid rumors of acquisitions and reprivatization, the company's stock fell by half in just a few months. Rackspace was forced to announce it was withdrawing from IaaS in May 2015, and would focus instead on services related to providing servers. Even Amazon began to feel the heat, and in their second quarter report, their CFO admitted that the price war had impacted their fiscal performance, and their stock took a hit.

By July, some overseas analysis companies and media outlets were discussing the possibility of Microsoft's cloud services' income surpassing Amazon's if they included SaaS.

The situation in the United States is already clear: the battleground the giants are fighting over is the public cloud. Only companies that have both financial resources and technology will be able to survive. Although Amazon's initial strategy was right by starting with IaaS, and that made it the long-term leader and gave it built-in advantages, as soon as Google and Microsoft entered the fray, who the ultimate winner will be became much more uncertain. Aside from the big three, who else will be able to get on board? Apple, Facebook, IBM, Oracle, Intel, Cisco, EMC/VMware ... The list of candidates is long, but the likelihood of victory is getting small.

What is the situation in China? In some part, it is very similar to a few years ago when Amazon was the undisputed leader and Google and Microsoft were still hesitating on the sidelines. In the Chinese market,

Aliyun is the undisputed leader. While Tencent Cloud has a comparatively complete product line, it seems to be in no hurry to grow, and Baidu's strategy has been unclear from the start, while telecom companies and other large IT firms don't yet have their heads in the game.

I once said, "Aliyun has no true rival in China. And the window will remain open for a rather long period time throughout 2013." In reality, Aliyun grabbed this chance and won some key victories as they continued to drive the narrative of the cloud computing industry.

Furthermore, Aliyun's income from Internet infrastructure and computing services broke $100 million. Although the company is still far from the size of its American counterparts, it has nonetheless ended the doubt about a "cloud bubble." Its nearly one million users make its future prospects seem bright. But the biggest breakthrough and turning point for Aliyun in 2013 was the overlooked engineering success of the Apsara 5K project.

Apsara is Aliyun's core system, with the ultimate goal of making tens of thousands of servers work as a single supercomputer and provide general computing services to the public. In 2012, Jian Wang said, "From a strategic perspective, what they [Aliyun] want to do could be interpreted as Amazon + Google and beyond." Making one computer cluster composed of thousands or more computers work is a sign of a country or enterprise's competitiveness. Aliyun had to overcome this technology challenge. It was just that Apsara, which began development in 2009, had never had the opportunity to reach the goal yet.

Apsara or Hadoop

In 2013, this opportunity arrived. During the first-quarter planning phase, they discovered that there were two data processing systems within Alibaba—the Hadoop (open-source) based Cloud Ladder 1 and the Apsara (home grown) based Cloud Ladder 2—both of which were rapidly expanding. It would quickly run out of storage space if the scale of the cluster couldn't reach more than five thousand computers across data centers. If they continued to invest in expanding both and build redundant capacity, the development and maintenance costs would be very high, and there would be massive waste. They had to stop working on one of the two, but which one?

There was a serious fight within the tech team over the answer to this question, and even Jack Ma was unable to stay above the fray. Hadoop was much more mature at that stage, given that it was the standard open-source big data project. It had the respect of the technology team, who were deeply committed to it. Hadoop also had bigger clusters to begin with. But issues of long-term development and security could potentially turn into long-term insurmountable obstacles. To address this, the Apsara 5K Project was launched with the ODPS (Open Data Processing Service) being upgraded into a five-thousand-computer cluster system, and development on Hadoop stopped and all of its programs migrated to the ODPS.

Apsara's core R&D team went through four months of hard work that involved optimizing the Pangu (storage system) and Fuxi (computing scheduling system) components while at the same time developing the Hua Tuo module (automatic diagnostic system). By August 15, 2013, the mission of upgrading Apsara was accomplished, and the ODPS cluster was successfully expanded to five thousand computers operating on the new Apsara 5K system. They also were able to run the whole system across different data centers and recover the system even when there was a power outage. Aliyun became one of a handful of companies in the world with this level of computing capacity, and also the first to make it available as a public service. It was one of the few times in the past few years that China has taken a leading role on the cutting edge of a key technology.

Apsara 5K is an important technology milestone in the history of Alibaba. Today, all of Alipay's data processing, Taobao's data banks, and AliFinance's loan services are supported by ODPS on Apsara 5K. These achievements have helped Aliyun prove itself within the group. After this, the debates about Aliyun, whether to use Hadoop or ODPS, and even about Jian Wang himself all ended. Apsara 5K put a period on these discussions. Aliyun became the group's single technology platform, something that everyone in the company agreed on. Recently, the thousand-plus HBase clusters have begun migrating to OTS (Online Table Service).

In October 2013, at the Aliyun Developers Conference, I watched the Apsara 5K memorial unveiled with many other Yunqi Alliance members. Jian Wang movingly expressed his gratitude for a long list of people who were still at Aliyun and many others who had left. Their names were engraved on the memorial. I know this memorial represents the culmination of Jian Wang and his team's work on developing their own cloud computing core technology, but their road extends far ahead. Because of his increasingly capable staff, Wang has been able to devote himself to strategic thinking about cloud computing and big data, and interacting with more clients, thereby having even a larger impact.

In reality, Jian Wang has always believed that whether Alibaba Group uses Aliyun services or not was not that important to Aliyun. This is because, if the company wants to become a general computing platform for everyone to use, the challenges it must overcome in terms of technology and service are far greater than internal use. Using Aliyun within Alibaba Group is just a side success.

This year, Wang began talking a lot with many different types of clients, and he was very moved. He said whether it is the government, finance companies, or small- and medium-sized enterprises, as soon as they move to the cloud, they are able to innovate far more than he imagined, something that often fills him with awe. Further, the clients' attitude toward cloud computing has already undergone a major shift, as more and more people are willing to accept this new technology. But with this many clients, are you able to meet all of their needs? The national railway ticketing website, for example, was not just an issue of government agencies but also involved the enterprises that work with it. When many problems remain unsolved, Chinese companies can't always blame the government; they must also take some responsibility themselves. Cloud computing firms need to boost their capabilities as quickly as possible; otherwise many clients will make poor decisions, such as taking the traditional tech framework and building their own data centers. "The thing I fear the most is that in five years a bunch of experts will call on the government to support the domestic cloud computing industry," Wang said, reiterating a frequently felt concern.

In this era, "The internet + data" has replaced "computers + software." Cloud computing is the platform and a prerequisite for making more industries and all of society digitized. Its value is not just that the platform itself makes money, but rather that it can serve as a form of infrastructure that allows the true value of data to be realized. This is far more valuable. In several recent conversations, Wang mentioned several cases of cloud computing users that he found moving and surprising. "Clients using Aliyun to do things is the main value of Aliyun." He said, according to data from the US Electrical Sciences Research Institute, if one iPad is charged every day, the entire cost for a whole year will be a dollar fifty, but who knows the value of all of the things people use them for?

In many respects, China's cloud computing industry could someday eclipse America's. Because companies like Alibaba have penetrated into many different industries, they might have a greater impact than their US counterparts, and because many domestic companies have incomplete IT systems, and Chinese society is accustomed to transformation, it is possible that China could leapfrog over an entire stage of development and build a completely new core IT system based on a cloud computing platform. For Aliyun and its partners, this is both a major opportunity and also an important responsibility.

Yunqi is an attempt to seize the historic opportunity of the internet in transition. Electricity was originally invented in Europe, but it first became a public utility in the United States, where it helped propel that country's twentieth-century industrial revolution. If we can soon make computing a public utility in China, we will be grabbing a huge opportunity provided by the internet economy era. This is a different opportunity for Yunqi and Silicon Valley.

At a forum of the Chinese Academy of Engineering, someone asked me, after the internet, cloud computing, and big data, what will the next big technology be? My answer was: For the next fifty years at least, all our work will revolve around the internet. In most cases, the biggest challenge we face are not ones in which everyone rushes in at once, but rather ones in which everyone gives up before they've figured out what's really going on, thereby losing many opportunities to grow and develop.

Chapter 8

The Fusion of the Internet,
Data, and Computing

Data has no inherent value, just as the footprints you leave on the ground have no value unless they are processed by computing power. Computing is like infrastructure: it can turn the accumulated grains of sand that are data into economic and industrial wealth.

The internet becomes infrastructure; data becomes the means of production; and computing becomes a public utility. Let computing be a power possessed by every human being and data an asset shared by the world.

The Internet Becomes Infrastructure and a Platform for Innovation

The internet has become the infrastructure for every country's economic and societal development, and it greatly expands the constraints of space and time in our daily lives. In what way is it a type of infrastructure? Simply put, it's just like highways and railways. However, the internet as infrastructure isn't bounded by the limitations of space and time as we know them, thereby surpassing roads and railways. Without infrastructure, a country has no GDP.

What is the internet? Today, people have vastly differing answers to this question. In the past, everyone was focused on the internet by itself, and now their attention has shifted to the mobile internet and industry internet, as people would like to call them. At the Global Mobile Internet Conference in 2011, when I was still very confused about its nature, I said that the mobile internet wasn't the same thing as the internet. Later, I discovered I'd been wrong. The mobile internet is still the internet. The mobile internet and the industrial internet, including what people refer to as the internet of things, are all part of the internet—their foundation is the internet. Today's internet is very different from that of ten or twenty years ago, inasmuch as it has become the infrastructure for economic growth and societal progress. Viewing it in this light is very important for understanding its true nature.

Over twenty years ago, the internet had yet to become a type of infrastructure. When I tried to open my first email account at the Institute of High Energy Physics of the Chinese Academy of Sciences, I had to request a favor from a friend. In order to receive emails, I paid six to seven thousand RMB to China Telecom to install a landline with a long-distance calling option. To dial up online, I had to make a long-distance call to Beijing. That was the only way to receive emails. You can imagine how much it cost just to get one email. Such an inconvenient and expensive internet obviously couldn't be considered infrastructure. But today is completely different. We can understand today's internet by studying its scale, costs, and convenience.

Scale is the first perspective that can be used to gauge infrastructure. Beijing's first highway was the Badaling Expressway (from downtown Beijing to the Great Wall). It was built to the highest standards of the time. However, in its infancy,

I wouldn't consider it a type of infrastructure. Regardless of how expensive or wide it was or how fast cars could drive on it, it was just a very convenient road that addressed the singular problem of traveling between Beijing and Badaling, a scenic area of the Great Wall. In 1972, when President Nixon came to China, he expressed his interest in seeing the Great Wall. However, a snowstorm struck Beijing on the day before the planned visit. The government mobilized hundreds of thousands of people to shovel the snow off the highway all the way from the Great Wall to Diaoyutai, the state guesthouse where Nixon was staying, in order to fulfill his wish to see the Great Wall. Today, the Badaling Expressway is a short part of the Beijing–Tibet Highway that stretches all the way to Lhasa. In this way, it has become part of a greater infrastructure. Similarly, the maglev that goes from Shanghai Pudong International Airport to downtown Shanghai cannot be considered a piece of infrastructure, while the high-speed railway system has long been an important part of China's infrastructure. Infrastructure has nothing to do with the technology behind a certain facility but rather is directly linked to its scale.

As our material standards of living have improved, people have started to realize more and more the value of highways. Without a highway, it is impossible to travel very far. In fact, it's hard to imagine living without such infrastructure. The whole point of becoming infrastructure is to reach a certain scale so that its full potential can be realized. Cloud computing is also like this. It must take the path of scaling up. Staying trapped within a small clique of users would be meaningless.

Cost is the second perspective that can be used to appraise infrastructure. Electricity is one of the most important elements of our modern infrastructure. Just think, if it weren't for environmental considerations, no one would think twice about adding a bunch of light fixtures, because the cost of power is no longer an issue; people are more concerned with the comfort of the lighting. But thirty years ago, almost every family in China would be anxious about the number of light fixtures in their house and the wattage of their light bulbs. At the time, electricity costs weren't something people could ignore. So, today when people discuss lowering the costs of broadband connectivity at home, they're actually talking about turning the internet into infrastructure.

The telegraph was once an important piece of the tele-communications infrastructure, but from the perspective of cost, it was actually quite expensive per word to send a tele-gram. When radio and television arrived, the cost per word of communication dropped dramatically. In the early days of email and text messages, the per-character cost of sending a message remained something that couldn't be ignored. At the turn of the twentieth century, the cost of a ten-character message sent across the ocean was about $100. In the early mobile phone era, the cost of a text message was 0.10 RMB (less than two cents), but today no one could accurately tell you how much a text message costs. Nowadays, you wouldn't think about the per-character cost when sending a message, because when the internet becomes infrastructure, the costs of communication are dramatically lowered.

Convenience or penetration rate is the third aspect we can use to evaluate infrastructure. When you're wiring electricity in a house, you must take into consideration the location of the power outlets so family members won't be worried about having no place to plug in. In other words, the convenience of electricity is very important. But now think about this: is it easier to find a power outlet or go online on the Web? Thanks to cell phones, it's easier to go online than to find a place to plug in. This is why many public places now offer free cell phone recharging stations, and why power banks become a necessity. One thing you may not have realized yet is that the internet has greatly outstripped all other forms of infrastructure in human history in terms of convenience. It is now a form of infrastructure that no one can live without.

What does that really mean? This is a brand-new challenge for all internet companies; it's the beginning of a new journey. For traditional companies, this is actually an opportunity that can allow them to stand at the same starting line as the established internet companies. This fresh start could reveal that an individual can have the same impact as a company with 10,000 employees. We don't need to worry about whether or not an individual still has the opportunity to launch her own start-up. From another perspective, what does this mean for a 10,000-employee company? With this new starting line, a 10,000-employee company can now have the agility to compete with companies of only one person.

When there is a new form of infrastructure, it's more important to consider first what you can use it for than what it can do right in general. When you build a road, your intention might be to enable the children in the village to get to school more easily. But there will certainly be other people who are able to use the road for other purposes. So I normally won't think about functionality now, but rather, I try to imagine what will be changed by this new infrastructure. That is the power of infrastructure. We don't have to argue about whether or not we have an "internet mentality." When someone starts to drive a car, they'll no longer maintain a "bicycle mentality." When they start to use electricity, they'll naturally develop the habit of paying electric bills. Infrastructure has to be a simple concept to grasp. That way people will be willing to turn a technology into infrastructure; otherwise no one would use it. We don't say that a truck driver needs to have a "highway mentality" in order to drive well. They're fine so long as they follow the traffic rules.

From the internet to the mobile internet to the internet of things, the evolution of those terms is really a reflection of the changes being brought about by this new emerging infrastructure. When people talk about the internet, they're talking about the traditional internet, i.e., the internet network that linked together all of the world's computers. When it linked up all of the world's smartphones, we entered the era of the mobile internet. And now people are talking about the internet of things, which will link together all of the world's physical objects. I like to call this the "internet of everything," and it will become yet another piece of infrastructure, only this time for all things. In other words, everything is being online.

As the internet became infrastructure, all companies and industries—whether they appeared to be directly related to the internet or not—went online. It's just like how, today, no one can say that life has nothing to do with bullet trains or highways. More than a hundred years ago, the railroads became a major part of America's infrastructure, bringing massive change to the whole country. The internet will have a similarly revolutionary impact on China and the rest of the world.

Once the internet becomes a type of infrastructure, we'll have an implacable dividing line in the sense that all the existing internet companies of today become simply "traditional" internet companies, and a whole new cycle of innovation will begin.

Data Becomes a Means of Production and a Type of Natural Resource

Data always existed—it's not as though it appeared only after the internet—but the internet made the sedimentation of data (the accretion of data like a sediment), and its use much easier and more natural and lowered its cost. This has offered our industries a historic opportunity for development. Today as long as there is a path, you will leave footprints. If your footprints can be computed, the offline services can be personalized based on what type of restaurants you went to, what type of stores you entered. It can even estimate your income and know what kind of person you're like just using the clicks of your mouse today. Why weren't footprints analyzed in the past? Because it was too costly to do so. Only police officers investigating a crime would measure footprints and analyze height and weight.

Viewed this way, data isn't something that is actively gathered; it is deposited as a by-product left by our activities, just like footprints. Footprints aren't actively gathered by the people who build roads. Rather, as long as there is a road, there will be footprints. It's impossible to build a road that would bear no footprints. There is no such road anywhere in the world. It's simple physics.

Data isn't a product of the internet era. It has been a part of society's fabric for the duration of human history. Why did "big data" become one of this era's buzzwords? Because the uniqueness of the internet as an infrastructure resulted in the faster accumulation of data. This has a lot to do with the technology of the internet: from TCP/IP protocols and internet routing protocols to the http protocol and hyperlinks, these are all the foundation for the natural depositing of data. As long as you "walk" on this infrastructure, you will leave "footprints." Therefore, data has become a true raw material, the first natural resource that humanity obtained on its own without having to rely on Mother Nature to provide it. Now what kind of impact will this new raw material have on the development of various industries?

In the future, everything on the internet will naturally be deposited as data, and as the cost of keeping those data goes down, we'll have the opportunity to analyze much of it. From your footprint, I can learn what type of shoes you wear, how much they cost, and by extrapolation your income, and I can do the same with data saved from the internet. Actually analyzing a real footprint is too much hassle, so no one bothers to do it. But with the internet of everything, that all changes. Data about walking will finally be recorded, and the accumulation of this data will become much easier. This proves once again that data is not collected—it is deposited.

The disappearance of Malaysia Airlines flight 370 in 2014 has a symbolic meaning from the perspective of the internet and data. On March 8, 2014, MH370 Boeing 777-200 lost contact with Ho Chi Minh City air traffic control on a flight from Kuala Lumpur to Beijing and disappeared from radar screens. Because of the events surrounding MH370, people learned that engine operation data could be sent to the airplane engine company without going through the plane's owner or maker. *New Scientist* magazine later reported that the Trent 800 engine used by the 777 in question was made by Rolls-Royce in Derby, England. The plant's engine health management center received two sets of data from the MH370 flight. It is impossible to conceive of this type of information transfer absent the infrastructure of the internet.

Today, engine manufacturing has turned into a data war. The Boeing 787 uses GEnx engines, which generate about five thousand data points per second, or half a terabyte of data per flight. Rolls-Royce's Trent engine has evolved from a device with dozens of sensors and hundreds of signals that would produce several megabytes of data into the Trent XWB, with hundreds of sensors, thousands of signals, and dozens of gigabytes of data points per flight. Pratt & Whitney is even more advanced with its PW1000G series—used on A320s and 737s—which boasts over five thousand sensors, a leap beyond the previous model that had less than one hundred. Each PW1000G can generate up to 10GB of data per second, but data are collected at a low speed. Even though they can only process a small portion of this data, it has still reduced the number of management personnel per engine from ten people down to one.

Right now, the challenges faced by airplane engine manufacturing may be ones of materials, craftsmanship, and design. But in twenty or thirty years, after these challenges are overcome, it doesn't mean anyone could just jump into the industry and start churning out quality engines. Why? Because a start-up would not have those twenty or thirty years of operation data. At that point, the core competitiveness of an airplane engine maker is more than just materials, design, and manufacturing. What MH370 tells us is that, with the arrival of the internet as infrastructure, data is a very important component of production and a new raw material.

The search for MH370 revealed that when traditional methods fail to locate a plane, only data can tell us the plane is still moving. But when there is only data left, would you believe it? Where is the airplane? There are so many information collection methods—radar, GPS stations, telecom communications—and none of them could tell us where the plane crashed. Yet data *can* tell you where the plane crashed, and it's calculated based on handshake signals between the plane and GPS satellites and the ping (a network diagnostic tool) signal, even though this was not originally designed for positioning. The first time this data was used to locate a plane was two or three years before MH370 disappeared. A plane flying from South America to Paris crashed in the Atlantic Ocean and could be found only by tracking its final pings.

From then on, people decided to use these pings for positioning and even added a time stamp, without which MH370 would not have been located. This is all beyond people's experience. I believe this is a classic case of traditional information not knowing what is going on in the world and data telling you the answer. At that point, the only question remaining is whether you're willing to believe it.

Again, data isn't gathered—it's deposited. What is purposely gathered is information, but what is naturally deposited is data. This is the unique quality of data as a raw material; it accumulates just as oil gradually accumulates. We are simply taking it out of the ground. We don't have the option of creating oil—it was created through a natural process and is a natural resource the Earth has left us over the course of hundreds of millions of years, whereas data is a natural resource that humanity itself has left. This is the first time in human history that we ourselves are creating a type of natural resource.

Data is an important strategic resource and at the same time a crucial component of products. As both raw material and natural resource, it requires legal protection. In the process of China's Reformation and Opening, guaranteeing land rights played a key role in boosting economic growth. By the same token, the question of who owns the data is fundamental. The lack of a clear understanding about who owns data will have a major long-term impact on the digital economy. Before the Reformation and Opening, the natural resource that is land faced the same challenge; it was resolved by setting farm output quotas for each household. The question for data today is the same one addressed when setting farm output quotas for each household: it is both simple and complex and will determine the liberation of the means of production.

Data is like oil and coal. We dig them out of the earth, but in reality they are a type of solar power that has undergone a billion-year-long process. The advent of the internet has transformed data into another natural resource, because the accumulation of data has become an exceedingly easy proposition. Blindly gathering data is not the same as a natural accumulation. Only the natural accumulation of data can be called a true natural resource.

Computing Becomes a Public Utility

Computing is a new public utility that is becoming part of each nation's and each enterprise's core competitiveness.

Computing made data accessible and therefore able to be used and to generate value. But using and flowing data requires a lot of computing power. Data doesn't inherently have value, just as a trail of footprints is worthless until it has been digitized and analyzed. But after your footprint has been computed, it becomes valuable. With all of the data on the internet, it would be impossible to analyze it all with just a few computers. You need thousands upon thousands of computers at a data center. At this point, computing becomes a public utility, and the form computing comes in is a cloud. Massive computing power has naturally become central to the competitiveness of enterprises.

From an industrial perspective, we can turn computing into a public utility just like electricity. The United States was the first country to turn electricity into a public utility, and this gave the country a massive advantage in promoting industrial growth. Therefore, if computing truly becomes a public utility, it would allow China to leapfrog more developed countries. Our historical mission is to launch the next round of economic development by being one of the first countries in the world to turn computing into a public utility.

Cloud computing as a public utility will kick off the next wave of the digital economy's development. Computing is as important to the digital economy as electricity is to the traditional economy.

As the internet's infrastructure further expands, two forces will push for computing to become a public utility. First, humanity's demand for computing is growing by leaps and bounds, and the appearance of data will only further increase this demand by orders of magnitude. Second, people now need to use computing anytime and anywhere, and therefore equal access to computing power will face a challenge.

Actually, before computers appeared, there was already a demand for computing, but back then a computer was simply a pen and paper. Writing down 2 + 2 = 4 is computing; it's just computing that is based on paper. Therefore the demand for computing was already there, but it needed the transistor to bring about the flourishing of computing power.

Transistors are the basic unit of computing devices. In fact, you could almost measure the demand for computing based on the number of transistors. The well-known Moore's Law (named after Gordon Moore's observation that the number of transistors in a dense integrated circuit computer chip doubles about every eighteen to twenty-four months) is calculated based on number of transistors. In 1971, Intel's 4004 CPU only had 2,300 transistors, but by 1979 when the first PC came out, the company's 8086 CPU had already increased the number of transistors over tenfold to 29,000. By 2016, Intel's CPU had more than 7 billion transistors. In June 2007, Edward Lazowska of the University of Washington made a shocking statement: in 2004, the world's production of transistors was equal to the annual output of grains of rice, if we match a grain of rice to one transistor. Our demand for transistors is just like our need for rice; it's just that most people have yet to realize this.

The iPhone 6 processor has more than 1.6 billion transistors. What does this mean? The Apollo spaceship's guidance computer had only about 12,300 transistors, meaning the iPhone 6 has 130,000 times as many transistors as Apollo. If you convert this into each computer's actual final processing capacity, it's even more startling. The iPhone 6's computing power is 120 million times as powerful as the Apollo's guidance system. In other words, the iPhone 6 could guide 120 million Apollo spacecraft.

On May 11, 1997, IBM's Deep Blue defeated chess grand master Garry Kasparov, marking the first time a computer defeated a grand master. Deep Blue is a supercomputer that ranked as the 279th most powerful computer in the world at that time. But Deep Blue's processing power is nothing compared to an iPhone. While we all carry around supercomputers in our hands, this is only one aspect of our computing demand in the internet era. When we use our cell phones to search, send emails, play games, or buy things, we're using the computing power provided by the cloud. It is hundreds or thousands of times more powerful than your phone. That is why you can no longer fulfill your computing needs with one computer you can buy. Now people need to have their computing demands met by computing power that is provided as a public utility.

Deep Blue weighed over a ton and was too expensive for a family or even small enterprises. Even if you had the same programming skills and idea, there is no way you could have made your own Deep Blue because the computing power was held by so few people at the time. At that time, cloud computing was not widespread, so this type of competition was restricted to large companies. That is the inequality of computing power access, which was much less equal than things were during the pencil and paper era. The goal of making computing a public utility is to redress that inequality.

In March 2016, AlphaGo defeated Lee Sedol, a level nine dan Go master from South Korea, in four out of five matches. When the news was announced, it once again sparked interest in artificial intelligence. Go's attraction lies in the fact that a nineteen-by-nineteen-square board can be host to permutations of moves—an astronomically high number of possibilities. In January 2016, *Nature* introduced a British AI firm called DeepMind, which developed AlphaGo, in a cover story. It explained the company's design principles and use of machine learning in detail. On the same day, it was announced AlphaGo defeated European Go champion Fan Hui, a level two dan master, 5:0. The matches had actually taken place in October 2015; the results had just not been announced yet.

Unlike a human competitor, AlphaGo took less than half a year to complete the jump from a two dan to a nine dan level Go player, mainly because of the way it learned. According to a report from the *Economist*, AlphaGo used 1,900 CPUs and 280 GPUs during its matches with Lee Sedol via the internet. While the article didn't explain their particular capabilities, people have extrapolated that AlphaGo's online system was the equivalent to approximately 25,000 to 30,000 Deep Blues. Before AlphaGo went online, it used incredible amounts of processing power to complete its initial learning. The machine started by studying 160,000 games played by masters, with a total of over 30 million moves, in just a three-week period. Then it played 1.3 million games against itself, a process that took 50 GPUs and around one day. AlphaGo then spent another week analyzing 1.5 billion moves with different strategies. All of these were just part of its comprehensive systemic learning. AlphaGo's difference with Deep Blue lay not only in its processing power, though. In 1997, an average programmer outside of IBM could by no means access computing power on par with Deep Blue. But today, if you have an idea and the programming skills, you can sign up for a cloud computing service like Aliyun and have access to the same amount of computing power used by AlphaGo, an amount tens of thousands of times greater than Deep Blue. This equitable access to computing power will help to create fair innovations.

In addition to the growing individual demand for computing power, another reason that explains processing power's ability to penetrate into so many aspects of our lives is that the unit cost of computing has steadily gone down. Something that cost $1 million to compute in the past may now be as cheap as a dollar. Meanwhile, computing abilities are constantly improving. Quantum computers are another massive leap of thousands or tens of thousands of times of computing power. One quantum computer is one to two million times more powerful than a Tianhe-2, which used to be the number one supercomputer in China.

Thanks to computing, our creativity and imagination no longer have to be limited. With a public infrastructure of computing, it won't matter whether you're running a company or are an individual, it won't matter whether you can afford to buy a supercomputer—computing will no longer be out of your reach. Consequently, I believe that once computing has become a public utility, everyone's creativity will be unleashed. Creativity is humanity's rarest resource; at the same time it is inexhaustible.

The Coming of the Computing Economy

From the perspective of development, the internet as infrastructure will be different from the traditional internet. When data becomes means of production, it is like the discovery of a new continent that will allow for complete sharing. This gives us a lot of room to think imaginatively about data, this new raw material. How will the economic value of data be created? The answer will be: by means of computing. That's why I refer to this new form of economy as the computing economy. Most of people would like to name it the digital economy instead.

Three key elements of the computing economy

The internet as infrastructure, data as a raw material, and computing as a public utility will combine to create a whole new economy. I call it the computing economy, and it is the economy of the internet era. The computing economy's main aspects and their interrelatedness can be shown in the chart on the facing page.

The computing economy doesn't mean the economy of today's existing internet companies, but its basic infrastructure is the internet. In the past, when people mentioned the internet, they thought this was just something that internet companies did and it didn't necessarily have anything to do with non-internet companies. The real economy isn't complemented by a virtual economy but by the digital economy instead. If the United States had treated the internet only as a virtual economy, it wouldn't have become such a major part of America's competitiveness. Without the internet, the United States couldn't fight a war in Afghanistan. US reliance on the internet as a basic part of infrastructure is already absolute; it's just that many people haven't realized it yet.

The computing economy's main aspects include being online. In the computing economy, if things are not online, there will be no potential economic value. The advent of the computing economy will unbind human creativity and imagination.

Data in itself has no value at all. It gets values through computing and exchanging. At this point, what is exchanged is the value of data, not really the property rights to the data.

Internet — interconnections — infrastructure

Being online — data — sharing and exchange — means of production — computing economy

Computing — public utility — power source

But the property right of data is a problem that needs immediate resolution. While data brings us new value, it also presents us with the problem of ownership. This problem is just like the land rights issues that China faced during Reformation and Opening, which caused the government to adopt policies that guaranteed a certain level of land rights, fixing the farm output quotas for each household, for example. If we don't clear up the issues surrounding data ownership, it will have a large impact on our future development.

On April 14, 2016, in Strasbourg, France, the European Commission formally adopted a new law, General Data Protection Regulation (GDPR), on the protection of data that replaced the 1995 European Data Protection Guidelines. The new rules brought several important changes.

In the old law, jurisdiction over legislation was granted according to country or region. But in GDPR, the regulation holds sway over any enterprise providing products or services to citizens of the European Union, regardless of where these companies' headquarters are located, and jurisdiction depends on where the data is being distributed. This dovetails well with the nature of the internet's flow of data, because the jurisdiction can flow along with the data.

Further, "the right to be forgotten" was also ensconced in this law. It clarified the right of data owners to delete individuals' data. If any users don't want their data to be used by an internet company, they can opt out and their data must be deleted. This emphasizes the individual's right to their data. This move is meant to establish the future of data in Europe and boost the EU's economy.

We can see that not just in China but throughout the whole world, when it comes to data, most people are still at the same place China was at the beginning of the Reformation and Opening. Today many companies still haven't accomplished what the Xiaogang Village farmers (the first village in China's Reformation and Opening) did. What did the Xiaogang villagers do? They disconnected the operation and ownership of land: the land was collectively owned, but the right to farm the land belonged to the farmers. Everything that was produced returned to the farmers. But GDPR on data protection didn't make clear a distinction between data property and use rights.

In the oil industry, crude oil is valuable, but it isn't the most valuable part of the industry. You can maximize crude's value only by processing it. If crude were the most valuable part, then Saudi Arabia would be more powerful than the United States. You can control oil production but not oil's value. Land can be sold for money, but its value lies in what is done with it. Humans' value is creating greater value on the same land.

The algorithm is the core of the computing economy

The algorithm demonstrates the great value humans contribute to the computing economy—it is the soul of the computing economy.

PageRank is one of the most famous algorithms. In 1998, Google founders Larry Page and Sergey Brin published a paper called, "The PageRank Citation Ranking: Bringing Order to the Web." It was about Google's early search system link analyzer and algorithm, which established Google's market position, allowing the company to achieve unprecedented success. The algorithm was key to creating the hundreds of billions of dollars of market value the company generates. This calculation methodology became a computing model that the academic world began paying more attention to as the company became more successful. The basic idea of PageRank is very simple. It analyzes the relationship between hyperlinks. A web page's importance grows depending on how many links there are to it, and its PageRank value increases. When a website with a high PageRank value links to another website, the linked website also goes up in value. The concept of a hyperlink first arose in 1945, but placing value on each link became the secret to Google's dominance over other search engines and became a core part of the company's value.

If PageRank's algorithm method influenced the internet in the past, then the blockchain algorithm will affect the internet in the future.

Most people first learned about the blockchain via Bitcoin, because it is the core technology of Bitcoin. But unlike with PageRank, the inventor of this technology is unknown. On November 1, 2008, a person calling himself Satoshi Nakamoto posted an article in which he explained his idea for a type of digital currency that would use what he called a chain of blocks, a technology that wouldn't be controlled by any third party organization. He wrote in his well-known paper, "Bitcoin: A Peer-to-Peer Electronic Cash System," "A purely peer-to-peer version of electronic cash would allow online payments to be sent directly from one party to another without going through a financial institution." Sometime in 2008, chain of blocks became block chain, and then blockchain.

After 2009, Bitcoin's market rose and fell dramatically, and its future remains as uncertain as the identity of its founder. But the blockchain itself has increased in importance. Some people believe that the blockchain will become as important and basic as the TCP/IP agreement. Everyone knows that without the TCP/IP agreement, today's internet would be very different. Can we imagine a future internet without the blockchain? If the blockchain becomes the basis of the coming computing economy, then regulation and oversight will be necessary.

Microsoft's Azure has also started to provide blockchain as a service. It has an encrypted, shared, and distributed ledger. This could be the first big new opportunity for Microsoft after having missed the boat on search in the early days of the internet.

In the future, our whole economic system probably will have to be redesigned based on these transformations. For example, managing imports and exports will evolve from import-export management "computers" to import-export management "computing." This in reality has already happened. The use of Aliyun's computing in America and Amazon Web Services in China is creating a flow of computing different from the flow of traditional import-export. Computing has already become an export in and of itself, creating "invisible but tangible" value.

In the computing economy, the nature of corporations will also change. In this setting, how will we evaluate start-ups? In the past, when we looked at start-ups, we would take three things into account: do they use computers? Do they design software? What kind of service do they provide? Today, the internet has become one large computer. Therefore, you have to look at whether a company has anything to do with data, and after that see how much computing power the start-up uses.

The internet's long-term impact on the Chinese economy has been greatly underestimated. The tragedy of Europe is in fact that the majority of its internet services are being provided by the United States, and there's no chance to change that. In the future computing economy, whichever company provides the most services to the world will monopolize valuable resources. The US internet population is only about 20 percent of the world's total, but American internet companies provide around 80 percent of the world's Web services. This competitiveness already has nothing to do with a company's market value. This is something people should not ignore.

Whether it is the European Industry 4.0 and the digital economy, or what the United States calls the sharing economy and the industrial internet, the end result is that they are the computing economy that is based on the fusion of the internet, data, and computing.

Chapter 9

From Moore's Law to the Laws of Being Online

Until now, our intelligence has been used to find ways to exploit nature, and now we must pay nature back for what we have taken. The computing economy will make this possible. It won't replace oil or make up for dwindling oil reserves, but rather it surpasses oil to become a brand-new natural resource.

Fire and Electricity

The online world is undiscovered territory that has yet to be opened up. The only limit to what we can do is our imaginations; nothing else is holding us back.

Today the internet, data, and computing are joining forces, and their combined impact on humanity has been drastically underestimated. This new force will continue to impact us for the next ten, twenty, even one hundred years. The pervasiveness of that change cannot be overstated and will likely far outstrip the work we're doing today. In the future, there will be a day when people won't be able to eat or find a place to live without going through the internet. Every aspect of our lives will be connected to it. So, everyone should think seriously about this. If anything in the history of human development compares to potential impact of the internet, data, and computing, it is fire and electricity.

The internet is like fire

In the early 2000s when I was talking with some friends, everyone was asking what the internet was and what value it would bring to humanity. I thought about this and said, if you look at it from the perspective of human history, then, to the best of my knowledge, the only thing that can compare with the internet is fire. Everyone is well acquainted with fire. It allows us to eat cooked foods and helps us expel the darkness. The impact of fire on humanity was massive.

Fire helped humankind conquer the unknown world, and it changed our brain cognition. The earliest turning points in human society all had to do with fire. The history of fire being used certainly goes back 500,000 years to when Peking Man in Beijing's Zhoukoudian was alive, and it may go back as far as one million or even 1.42 million years ago, when the earliest human ancestors arose in Africa. Humans struck flint stones to create sparks for fire, and from then on greatly increased their understanding of fire until, ultimately, we learned literally how to hold fire, this glorious object, in our hands. Fire can be used for hunting, making weapons, making food easier to digest, keeping out the cold, bringing in the light, making containers, opening up new lands for farming, smelting metals, and many more things.

Computing as electricity

When electricity became a public utility, the American continent completely transformed. Why was electricity, which was discovered in Europe, first used to launch industry in the United States? One of the most important steps in the process was when the United States turned electricity into a public

utility. Before that, as I noted earlier, you had to buy a generator or build your own power plant in order to have electricity.
In the late nineteenth and early twentieth century in the United
States, if you were going to build a factory, you also had to
build a power plant. If the factory went out of business, the
power plant would be shuttered. The power plant was often
the highest cost of running a factory. One of Edison's greatest
contributions was transforming electricity into a public utility
in the state of New York, and the United States became the
first country in the world to make electricity a public utility.
Electricity's transformation into a public utility became one of
the major driving forces in America's development.

With the internet, the value of computing has been rediscovered. Computing isn't brand new, and it isn't something
that came into existence only because of computers. When
people used knots on ropes to count or paper and writing utensils or slide rules to make calculations, they were
computing. Compared with knots in a rope or an abacus, the
computers we use today are truly amazing. Everybody saw
the equipment was progressing yet couldn't imagine such feats
of computing power were coming and would be of such great
benefit to humanity.

Actually, we have already passed the period when people
are focused on costs, which is not to say that the expense of
computing isn't important; it's just that the problems of cost
have been resolved by technology. At this time, we're finally
able to look at the value computing generates.

Being online allows us to see a complete world
Our understanding of the world progresses amid contradic-
tions. From physics, we understand the dual particle and wave
nature of light. This problem—of the nature of light—has been
addressed by countless ingenious physicists over centuries.

In 1655, at Italy's University of Bologna, Francesco Grimaldi observed that light moves in waves like water. In 1672, Newton used light diffusion experiments to discover that light was made of many different colors. He argued that light was a particle. In 1921, Einstein won the Nobel Prize for his theory of the dual nature of light, showing that it was both particle and wave.

Being online will allow us to form a more complete picture of the whole world. When you understand being online, you will discover that in addition to all the offline knowledge we have acquired, there will be a whole new world of things to discover, just as our minds expand our capacity for understanding outside the physical constraints of our bodies. Being online will give us a more complete world. This is something I firmly believe.

Being online won't be limited by the offline world. Being online will liberate us from the traditional constraints of the physical world.

In the past, people used the term the "virtual" world to describe the internet, implying that the online world was somehow different than the world that everyone was used to. People didn't realize the appeal of the online world and how big of an impact it would have on our future. No one thought long and hard about the nature of being online.

In the offline world, you and I live in the same time and space. Every time you walk down a road, it's a real thing. If you and I meet up, it is definitely happening at the same time; there is no way that I am in a daytime and you are in a nighttime. But in the online society, we can have an interaction not constrained by time and distance. In this regard, the online world is an additional new world. Being online changes the time and space of human existence.

Google and other companies are promoting online translation. In May 2013, Google's annual I/O conference was held in San Francisco. There, a Google engineer released data that Google had never shared before: Google Translate provides 200 million people with responses to 1 billion requests every day, 92 percent of which are outside the United States. At the time, Google Translate could translate between seventy-one languages and had recently added Farsi, Hmong, Cebuano, and Javanese. Nonetheless, this is still only a small fraction of the thousands of languages in the world. This is a service that could only be provided online because its scale is too large to be handled by an offline system. After all, Google Translate translates one million books a day, a number greater than the output of all the world's professional translators combined. There is no way the offline world could produce that much work.

Being online has liberated us from such limits and given us a whole new world.

This is why, if you want to draw an historical analogy for computing, the steam engine also comes to mind. The utilization of fire, the discovery of the Americas, and the invention of the steam engine all took many thousands of years to occur. I believe we live in an era when fire is being mastered and the steam engine is being invented. Their equivalents are the internet, data, and computing. This should be a very exciting and hopeful era.

The online world is an unexplored land. Our work hasn't yet begun, and there are still many things that can be done. The only limit in the online world is our imagination, nothing else.

The Laws of Being Online

The laws of being online can help you understand whether something belongs in the online world.

The new world will have new rules of the game and new laws. Being online is restructuring the basic givens of our lives. Seemingly minor changes are creating whole new rules and products. What the online world will become depends entirely on our imaginations. We exist in this online world, but it can develop only if we understand its true nature and learn how to use it to protect ourselves. Several years from now, when we've gotten used to the state of being online, no one will focus on discussing this issue, just as today no one asks why the sun is hanging above our heads.

But today isn't like that yet. I would like to ask a question: Do you think you've met more of your friends in person or not? By "met in person," I mean met offline. By "not," I mean have contact only through the internet.

Research shows that the number of friendships maintained online is roughly the same as the number of friendships maintained in person. University of Oxford anthropologist Robin Dunbar proposed the rule of 150 based on research into people's habits of sending Christmas cards. The rule says that most humans maintain a stable social network of around only 150 people. Facebook's analysis has revealed a similar result. Online, we normally maintain active relationships with only a similar number of people.

Actually, the relative number of friends and acquaintances offline and online is probably not significant. The important thing to realize is that when the number of our online relationships makes up half of the total of our relationships in total, our lives will experience a complete transformation.

The American marketing guru Joe Girard has put forward the law of 250, which says behind every client there is a network of 250 friends, relatives, and neighbors. This means that if you can establish a good reputation with one customer, then potentially you will enjoy the favor of 250 people in that client's social network. On the other hand, we know that on the internet offending one person can result in your offending 100 million. The difference in numbers is striking and significant. That is the difference between being online and being offline.

Online means you can drastically reduce the frequency of your trips and the number of times you'll have to take a train or airplane. The proportion of your expenses taken up by the internet and online devices will continually rise. Good delivery by online takeout services will be more meaningful to you than eating at a fancy French restaurant. Working from home will no longer be unusual; meetings will be through online videoconferences. The COVID-19 pandemic that has forced so many to work from home, restricted social gatherings, and prevented dining out has given us a foretaste of all this, and has shown how humans can adapt to being online in the future. As an advertising slogan for online shopping says: just because there's no one in the streets doesn't mean no one is shopping. The video accompanying the slogan shows an abandoned road, empty coffee shops, cars all parked on the side of the road, while the most fashionable clothes in the world are being purchased online.

The way I see it, being online will bring a whole new type of social relationships, including new kinds of jobs, organizations, families, marriages, friendship, and consumption. The reason is both mysterious and simple: because they will all be linked to the internet.

In the digital era, Moore's Law was very important to understanding the future of the whole industry, not just semiconductors. Today we have entered the online world. So what's the law of being online? We need a different approach to understand this world. There are three laws of being online that the internet, data, and computing have brought to us.

Law one: Every bit is online

You could say that everything in the world is in its atom state, or in other words, in its bit state. We can understand bits by their physical counterpart, but actually, that definition is unimportant. A bit's ultimate state is that it must be online. In this way, everything in the world will be linked together by their bits, and all bits are accessible.

Even something like a tree can be given a marker by humans that could then be connected to an online system. Based on the online system's development, atoms and bits will become the state of the future.

The IPv6 protocol resulted in addresses that are 128 digits long. Compared to IPv4's 32 digits, there are now enough permutations to give every grain of sand in the world an IP address. It seems as though this new protocol was made with the emerging online world in mind.

Law two: Every bit must flow on the internet

If the offline world had no solar, fossil fuel, and wind energy, it simply couldn't function. Data is an energy that can give life to the whole online world; it is the foundation of the system's operation. Every bit represents data that must flow freely—that is the only way data can be the source of energy for the online system.

If the online system doesn't have free-flowing data going over it, then it will simply revert into an offline system. As soon as data stops flowing, it is just as if you and another person stop communicating with each other, and the system will soon go offline. I want to emphasize that the online system isn't considered online just because something is connected to the internet. It is the online system only if data is flowing freely over it.

It's like a road. It doesn't matter if you cover it with tar. If no one uses it for ten years, weeds will still sprout. Devolution is something that always happens without people noticing it.

The reason I'm opposed to the private cloud is in essence because the private cloud doesn't let data flow freely over it, and the data circulates only within its own system. People must understand that in an opaque and closed system, data isn't serving its natural purpose, which is to meet the needs of society. The private cloud is the America Online model of the internet. But today we don't refer to the internet as AOL.

The true movement of data must be over the internet, not on local area networks. Sometimes I say to people that, if you want to talk about big data, then remember that "big" refers to the big *scale* of the internet. Locking things up within a company has no future. Data that flows over the internet is far more valuable than data on a local area network. The logic is obvious, but that doesn't mean everyone understands the importance of this fact.

LANs and the internet are very easy to differentiate. When AOL was at the height of its popularity, I was in the United States, and when they advertised their website, you had to get a disk and use the disk to install specific software on your computer before you could use AOL. That was in order to exclude some users and include others. Although AOL offered every type of service—email, news, forums, etc.—it was actually a closed network. It's just that at the time, it was the world's largest closed network, so it seemed like it was the entire internet. When the open internet became widely accessible, AOL's closed model became unsustainable.

As soon as data moved freely, the value of the information on the traditional system grew exponentially.

Law three: Every object represented by bits is computable on the internet

In 2006, MIT professor Seth Lloyd wrote in his book *Programming the Universe*, "What does the universe compute? It computes itself. The universe computes its own behavior. As soon as the universe began, it began computing. At first, the patterns it produced were simple, comprising elementary particles and establishing the fundamental laws of physics. In time, as it processed more and more information, the universe spun out ever more intricate and complex patterns, including galaxies, stars, and planets. Life, language, human beings, society, culture—all owe their existence to the intrinsic ability of matter and energy to process information." The world was born in the Big Bang, but whether the old offline world can be computed is something that awaits research. The online world, which is composed of bits, should be able to be computed. You could say that computing is a natural characteristic of the online world.

The mobile internet has precipitated another transformation of the internet, because it closely adheres to the three laws of being online. First, the mobile internet is reliant on portable devices being online. Second, the mobile internet can be online anytime and anywhere, so you can't use the offline mindset to manage or operate it. Third, because you can be online at any time, that means data is moving at any time. The mobile internet is currently the space with the highest rate of exchange of data. This adherence to the three laws has made the mobile internet the industry with the most hope and most vitality.

The internet of everything is even more representative of this trend.

After everything is online, cutting-edge technologies of the past will be able to be integrated. The hottest technologies of today were actually being discussed decades ago, including the smartphone, wearables, AI, virtual reality, and augmented reality. It was once thought impossible to develop them in concert, but being online has breathed new life into them all at once. Wearables and computing are often linked together when discussing implanting chips inside the human body and recording people's every movement. In this framework, you should know by now that, if the chip isn't online, then it's worthless. Similar technologies will be redefined by being online.

In 1853, *Scientific American* wrote, "Up until 1846, our nation still did not have a single article of clothing that was made with a sewing machine; in that year the first patented sewing machine was made. Today, tens of thousands of people are wearing clothes made by sewing machines, and every article of clothing is on par with some hand-made clothes." Similarly, when it comes to online technologies, perhaps most people haven't been able to develop a clear impression. But when the products reach the market, they'll be quickly adopted because that's the major trend of this era.

We need to take a new approach to explore this new world. The new approach means using the cloud and data.

The cloud approach requires us first to look at its scale when we are evaluating something. This type of scale is hard to imagine in the offline world. What needs to be slowly learned is enlarging the scale 10,000 times. For example, if something you may have had to spend three years to complete is done in a day, you will unconsciously start searching for something else to do. This is certain. Before there were airplanes, a trip from Hangzhou to San Francisco and back could be taken only a handful of times in a lifetime. There have been a few years when I've made two round trips between Hangzhou and San Francisco within seventy-two hours. Scale has changed a lot!

By this logic of scale, you can think about all the things around you that aren't online and imagine what they'll be like after they go online.

Using data to think about issues isn't the same as traditional statistics. Society is changing from one of individual participation to one of data participation. Whether you like it or not, data decides many things. Today, the news you read isn't edited for you, it is data-selected for you. The "You might also like this . . ." link long ago became a function that is de rigueur for shopping, music, and news websites.

If this link were not designed by data but rather relied on surveying you, the expense would be very high. In reality, it doesn't matter how much effort you spend, you couldn't replicate this function with manpower. This is something that only a data mindset could produce.

There is a line from the DreamWorks movie *Kung Fu Panda*: "Yesterday is history; tomorrow is a mystery; but today is a gift." This line tells us to cherish the present moment. If we just had enough data, it could explain the past and clear up a lot of the future's mystery. We can gather more and more data, and we would learn much more from all of this data. The cloud and data combined will create a chemical reaction to form cloud data that greatly exceeds the value of each on their own. In the online world, there is no big data, only cloud data, only free-flowing data. That's the only data that has value.

You can imagine many things around us that will change thanks to being online. Everything in the world will be online. I believe that only 1 percent of what can be optimized through being online has actually been put into practice today. That means 99 percent of the world is left to be put online. I hope that in the future we can put more effort into doing a better job of being online.

The offline world has had civilization for around 6,500 years. As of today, we are in the first decade of the next 6,500 years of online civilization, and we've already had significant changes. We have already reinvented assets of the offline world, as with Google Glass. But many things that truly belong to online society have yet to arrive.

Once, I ran into a friend who works in Lishui, a remote city far from Hangzhou in Zhejiang province. I asked him a stupid question: how could a city as remote as Lishui produce the Longquan sword and Longquan celadon, both of which became famous around the world? In my understanding, a closed-off region couldn't have produced such important artifacts of civilization, particularly in ancient times. The Longquan sword is one of the ten most famous swords of ancient China. Its craftsmanship is of the highest quality. Further, the workshops that produced the celadon china were among the five most important of the Song dynasty, and the blue color of their products is famous the world over.

My friend gave me a simple explanation: There is a river called the Oujiang that flows through Lishui and is like a superhighway connecting it with the outside world. Civilization relies on movement, and the faster the movement, the more developed the civilization. The Oujiang River was a conduit of civilization for Lishui. In human history, the four greatest ancient civilizations all relied on proximity to waterways. Being online is like the Oujiang River. No, it's not just the Oujiang. Its impact on civilization will be far greater than that of a single river.

Offline, we evolved from primitive humans by adopting the use of tools, walking upright, mastering fire, and ultimately developing our large brains. Being online has provided us with a whole new tool, and because of it, we are facing another leap of evolution. Are our brains ready for it?

Shadows in the Sunshine

In the online world, to get what you've never had before, you'll have to pay a price you've never paid before.

The shock to the system that being online will bring is of course going to be extremely large. Being online changes so many things and in the future will change even more. This is an incontrovertible fact. But what being online means for society is something that needs further discussion.

In 2008, the *Atlantic Monthly* published an article by Nicholas Carr entitled "Is Google Making Us Stupid? What the Internet Is Doing to Our Brains." Carr observed:

> Over the past few years I've had an uncomfortable sense that someone, or something, has been tinkering with my brain, remapping the neural circuitry, reprogramming the memory. My mind isn't going—so far as I can tell—but it's changing. I'm not thinking the way I used to think. I can feel it most strongly when I'm reading. Immersing myself in a book or a lengthy article used to be easy. My mind would get caught up in the narrative or the turns of the argument, and I'd spend hours strolling through long stretches of prose. That's rarely the case anymore. Now my concentration often starts to drift after two or three pages. I get fidgety, lose the thread, begin looking for something else to do. I feel as if I'm always dragging my wayward brain back to the text. The deep reading that used to come naturally has become a struggle.

Carr went on to write the 2010 book *The Shallows*, where he used cognitive and neuroscientific research to prove that the internet was negatively impacting our brains by impairing our ability to concentrate and think deeply. But in the *New York Times*, the reviewer likened Carr's concerns to the fears Socrates had toward books: "Socrates started what may have been the first technology scare. In the 'Phaedrus,' he lamented the invention of books, which 'create forgetfulness' in the soul. Instead of remembering for themselves, Socrates warned, new readers were blindly trusting in 'external written characters.' The library was ruining the mind."

Many profound impacts of an event are not discovered until long after it has happened. We have only recently realized that thanks to light bulbs, we have more evening activities to do and as a result sleep less. The arrival of light bulbs altered our bodily rhythms, and productivity changed. Suddenly, we got used to nightlife consumerism, and at the same time extended work hours spurred the faster development of society.

Fire was once the cohesive spirit of the family. Family members would gather together in the communal room around a fireplace and talk about their day or other matters. With the arrival of electric lighting, people were given another option: they could retire to the privacy of their own room and read, work, or do something entertaining. Electric lighting, therefore, loosened the cohesive bonds of the family.

That lighting would have such an impact on society was something no one would have ever guessed at the time.

The internet of everything is emerging, and I believe we need to be even more cautious than Carr is.

Being online will reveal all of our deficiencies. For example, if airplanes and their engines are online, people on the ground with malign intentions will have only to access equipment connected to the same network to cause the airplane to lose control and crash. The cyber threat is real. This "new security" concept will force all related technologies to change. Google and Tesla launching autonomous cars means that the world of the unknown has already begun.

Other problems have arisen. When driverless cars are everywhere, who will be responsible when two of them crash into each other? Is it you or the other car owner? Is it the car company or the map provider? Who is at fault? How will the policies of car insurance companies have to be rewritten in order to account for these new conditions?

Being online has broken the limitations of time and space imposed by the offline world, and this requires new rules. If you try to use the old offline rules to judge events happening online today, you'll fail. That is inevitable. Whenever you discover the same rule being applied both offline and online, there is most likely going to be a problem.

After moving for fifteen years in the direction of the internet utopia, Wikipedia finally discovered its biggest challenges: maintaining entries' neutrality and verifiability. Under the banner of a free and open system, people have filled Wikipedia with spurious content—sometimes inaccurate or biased or fluff—and the site is essentially under siege by these types of contributors. This isn't Wikipedia's fault; it's just that the rules of the game have changed. Just as in the offline world, we must continue to work to improve the rules of the online world.

In the second season of the television show *Homeland*, there is a plotline that revolves around terrorists getting a hold of a code for the vice president's pacemaker. They use a remote device to overload the pacemaker and trigger a heart attack right in the vice president's office. This may seem like something that could happen only in a TV show, but it could definitely happen in real life, and that would be even more dramatic.

On July 25, 2013, at the annual DEF CON hacking conference in Las Vegas, it was announced that rock-star hacker Barnaby Jack was found dead in his home in San Francisco. Jack had revealed a way to hack ATM machines so they would disgorge reams of cash. In 2012, he discovered a way to attack an insulin pump remotely. At the 2013 DEF CON, he was planning to unveil a new hacking technique that would allow you to access pacemakers and other medical devices from thirty feet away, install different software, and launch attacks. Unfortunately, this white-hat hacker died young.

In the past, there was a saying: on the internet, no one knows who you are. Now there's a new saying: on the internet, everyone knows who you are. This is a classic example of going from unknown to known. Privacy issues have become a hot topic discussed at every level of society.

Time magazine's person of the year in 2006 was "you"— meaning everyone using the internet. Announcing the award, Lev Grossman wrote this:

> You—yes, you—are *Time's* person of the year.
>
> It's a story about community and collaboration on a scale never seen before. It's about the cosmic compendium of knowledge Wikipedia and the million-channel people's network YouTube and the online metropolis MySpace. It's about the many wresting power from the few and helping one another for nothing and how that will not only change the world, but also change the way the world changes.

Citizens of the internet becoming person of the year is a milestone, coming only twenty-four years after *Time* named the personal computer "machine of the year." Everyone on the earth can claim they were the person of year 2006.

But the perspective from which people view things is always changing.

When I was writing my dissertation, putting a URL into your bibliography was a cause for concern because you never knew when that Web page might disappear. Therefore, I preferred to go to the library and consult with staff to find a book I could cite rather than simply input a URL. Today, no one would do such a silly thing, much less imagine that one day the internet would simply stop working.

Now people are swaying between confidence in and fear of the online world. This type of constantly oscillating attitude and judgment is gradually changing some very basic things.

Several years ago, Tesla was criticized for its vulnerability to hackers due to what people thought was an insecure API identity verification. Tesla's API allowed drivers to use iPhones or Android phones to link up with the car's computer to check the battery status, control the thermostat, raise and lower sunshades, recognize the car's location, honk the horn, open the charging port, and even record the car's movements. The point of this story is that before Tesla put its cars' API online, no one had ever thought about the possibility of the car being hacked. The criticism shows that people are now using an online mindset to look at problems. The car is just one example of huge challenges we face. When everything goes online, the challenges include large-scale software systems with security vulnerabilities, large-scale data with unknown security and privacy risks, and complicated algorithms that no human can understand well. Are we ready for these challenges?

Being online can speed up commerce, innovate new services, and improve society. In the beginning the internet arrived like a flood breaking through a dam. Only now are people learning to control the raging waters. Our exploration of the online world has yet to truly begin, meaning our understanding of its long-term impact on humans is even farther off.

A middle school teacher I know told me that without the fountain pen, human civilization wouldn't have progressed to where it is today. In China, calligraphy brushes and their accoutrements—ink, table, paper, et cetera—were too expensive for most people. Only a wealthy family could afford to use the old brushes. There was no way these writing utensils would have become widespread and gained the popularity among everyone. It was only when the fountain pen was invented that we saw widespread increases in literacy and writing ability. But this teacher added that the ability to write calligraphy was nonetheless considered a sign of high culture. Online civilization too will have situations similar to the transition from brushes to pens.

In nineteenth-century England, trains were a symbol of a world out of control. George Stephenson's first passenger train from Darlington to Stockton could travel at the incredible speed of fifteen miles per hour. Trains devoured villages in their path, destroyed the middle class, and sometimes crushed people to death. Cities, too, have often been demonized. When cities first began growing in what became Germany, there was a saying: "Before you go to Munich, make sure you've written your last will and testament." But today no one would deny that cities are one of the most amazing achievements of humanity. In fact, many believe the city is the greatest invention of human beings.

Chapter 10

Humanity's Self-Confidence

To have respect for others is to have confidence in ourselves. Data and computing have given humankind previously unimaginable capabilities yet have also resulted in unprecedented uncertainties that now challenge humanity's confidence.

Showing Respect for Others, Being Confident in Oneself

This is the greatest era of any time. Being online has given us incredible opportunities that are without equal in the history of the human race. What kinds of changes has the internet, which has gradually become an infrastructure, brought for the innovators of this generation? Thanks to the remarkable ease with which it can spread, the internet has become the foundation of social and economic development today. The combination of data and computing has unleashed enormous power that we have never wielded before, but with it also come unprecedented uncertainties that are challenging our self-confidence.

After speaking about the internet, data, and computing, we need to consider what we can do for technology development over the next fifty years. Any prediction for the future would be impossible to guarantee, yet three events that transpired during my lifetime shed some light on the nature of the challenges ahead: the first time people summited Mount Everest, the Apollo program to put a man on the moon, and the birth of the Boeing 747. These three events not only left deep impressions on me but changed my attitude toward life.

On top of a mountain

The height of Mount Everest is said by China to be 8,844 meters, by Nepal 8,848 meters, and by the United States to be 8,850 meters. Regardless of which metric you use, it is the tallest mountain in the world, a miracle created by nature, the conquering of which has become a dream of many.

At 11:30 in the morning, on May 29, 1953, for the first time ever, humankind reached the top of Mount Everest, a feat achieved by Edmund Percival Hillary of New Zealand and Tenzing Norgay, a Nepalese guide. Even though the two of them stayed on the summit for a mere fifteen minutes, they left an enormous gift for humanity.

Norgay described the situation at that time in the book *The Tiger of the Snows*: "A little below the summit Hillary and I stopped. We looked up. Then we went on. The rope that joined us was thirty feet long, but I held most of it in loops in my hand, so that there was only about six feet between us. I was not thinking of 'first' and 'second.' I did not say to myself, 'There is a golden apple up there. I will push Hillary aside and run for it.' We went on slowly, steadily. And then we were there. Hillary stopped on top first. And I stepped up after him."

This was Norgay's seventh climb, but the person who went up first was Hillary.

If you read news reports from the time, you will see that the person in the picture of the first to climb Mount Everest is Norgay, the guide, instead of Hillary. Norgay didn't know how to take photos, so Hillary asked him to stand on the peak, and took a picture to send out to the whole world. The world saw Norgay with ice pick in hand. This is not a picture taken for a single person but a picture for all of mankind. When Norgay asked to take a picture of Hillary, Hillary just shook his head and refused, so he left without a record of himself reaching the mountaintop, but that didn't matter: his image was still imprinted on the five-dollar coin of New Zealand. Hillary climbed to the top first, but he took a picture of the guide on the mountaintop instead. This is something that few of us would do. Later on, Norgay was considered a "good" guide, something that needed no proof, given the fact he was one of the first to reach the top of the mountain.

The real organizer of this climbing adventure was Colonel John Hunt, who led an expedition team of more than four hundred people, including 362 porters, twenty Sherpas, and more than ten thousand pounds of luggage. On May 26, Tom Bourdillon and Charles Evans made the first attempt on the ascent but weren't successful. The vertical distance between them and the summit was only ninety-one meters. Colonel Hunt selected Hillary and Norgay for the second attempt, and compared to Bourdillon and Evans, they got lucky.

Had it not been for Hunt's vision, the mountaineering party wouldn't have been there. Had it not been for Hillary's determination, Norgay wouldn't have been there as a guide. Had it not been for the guide, Hillary would never have been able to get to the summit and view the unbelievable vista.

But Hunt's expedition team wasn't the only one with such vision and determination.

In 1921, 1922, and 1924, Englishman George Mallory made three attempts to summit Mount Everest. Though unsuccessful, he became someone who would be always remembered and respected. Mallory and John Maynard Keynes, the famous economist, were classmates and good friends. On June 8, 1924, Mallory and Andrew Irvine made a final attempt together. On their ascent, they were covered by a cloud eight hundred feet from the summit, and that was the last either was ever seen until April 30, 1999, when an American expedition team discovered Mallory's body, which was mummified at around 8,170 meters. By the standards of modern technology, Mallory's climbing equipment was quite primitive, and even the oxygen he carried wasn't sufficient for him to ascend to the top. However, his wife's photograph wasn't found on Mallory's remains. He once said that he would leave the picture at the top of the mountain if he were to summit successfully, so I would rather believe that Mallory was the first person to reach the top. Some people have said that the remains of these dream-chasing mountaineers were the road map and ascending to the top was only the beginning, while coming back alive was the ultimate goal.

Summiting Mount Everest wasn't done to prove the climbing capability of humankind. As Hillary once said, "It was not the mountain that we conquered, but ourselves." However, the mountains have inspired the development of humankind's capabilities and stretched the limits of our imagination.

Into space
The only thing higher than the mountain is the sky.

The first human space capsule, *Vostok 1*, was launched by the Soviet Union on April 12, 1961, achieving escape velocity for the first time ever to send astronaut Yuri Gagarin into space. On May 5, 1961, the United States was also able to send astronaut Alan Shepard, in his *Mercury 7* capsule, into space. Shepard was just twenty-three days behind Gagarin; however, Gagarin achieved orbit and circled the Earth, while Shepard flew at a suborbital trajectory and failed to circle the Earth. Gagarin's *Vostok 1* spacecraft weighed 10,428 pounds, while the *Mercury 7* weighed 2,100 pounds. Gagarin had been weightless for eighty-nine minutes, while Shepard's weightlessness lasted only five minutes. The apparent disparities in technical competence raised a question about the true strength of the United States.

On May 25, 1961, President John F. Kennedy announced to the whole nation the objective of landing an American on the moon and returning safely to earth before the end of the decade. This was the Apollo Project, which is comparable in scope only to the building of the Panama Canal in American history. The Apollo Project started in 1963 and ended in 1972, as the head-to-head competition between the United States and the Soviet Union had become a technology race. At 9:32 a.m. EDT on July 16, 1969, *Apollo 11* took off from Kennedy Space Center at Launch Complex 39. The spacecraft reached the moon on July 20, and the lunar module landed at 4:18 p.m. EDT. Armstrong became the first man to step onto the lunar surface on July 21 at 10:56 p.m. EDT. The first words that he said were, "One small step for a man, one giant leap for mankind."

If you were to read news reports from then, you would find that, while it was an American who landed on the moon for the very first time, what was written in the headlines wasn't that the moon was landed on by an American but by all mankind. At the time, this was indeed the only successful landing of humans on the moon. The project started during the Cold War with the goal to inspire the confidence of the American people. Amazingly, the scientists and the engineers were able to safely accomplish the mission before Kennedy's deadline. The most extraordinary part about the Apollo Project was the display of confidence. "Ask not what your country can do for you—ask what you can do for your country," Kennedy had declared. That is exactly the type of spirit the people working on the project possessed.

That confidence was displayed for the world when *Vostok 1* and *Mercury 7* were launched. The world didn't know about Gagarin's space travel until after the successful launch and return of the spacecraft, while Shepard's short trip was broadcast live to millions of viewers. From that point on, the United States made every single launch a live event for the whole world. Live broadcast of the launches might expose all your defects to the public, yet with the confidence to be honest about your problems, it's just a matter of time before you improve your technology and succeed.

Such confidence is more important than anything else. To this day, the Apollo Project remains the greatest display of leadership in United States history. The Apollo Project laid the foundation for many of the technologies we rely on today in areas ranging from communications to biomedical technologies and advanced materials and spurred innovations in many other industries.

On July 3, 1969, at the Soviet Union's secret Baikonur launch site, the N1 rocket exploded when it was still attached to the tower and destroyed the launch pad. The N1 resulted in the largest artificial nonnuclear explosion in history, the equivalent of 7,000 tons of explosives, about the same scale of the Hiroshima bomb. It became known to the world only through the spy satellites of the United States. After another incident with the N1 rocket in November 1972, the curtain finally fell on the USSR's moon landing project. But the Soviet government didn't officially announce the end of the project until 1989.

On April 7, 2011, during the sixty-fifth session of the United Nations General Assembly, it was declared that April 12 would be the International Day of Human Space Flight, commemorating the day when Yuri Gagarin flew into space for the very first time on *Vostok 1*.

Since then, humans have gone into space many times, yet it is still not possible to go without high risk. People have stepped on the moon, yet the moon is so far away from us still.

Up in the sky

Between outer space and the mountains is the sky, within which we're able to move freely today. We have someone to thank for that—Joe Sutter, chief engineer of Boeing's 747 program.

On February 9, 1969, the Boeing 747-100 jumbo jet took off for the very first time, several months before Apollo 11's successful landing on the moon. While the whole world was cheering for the successful moon landing, Boeing was more than $2 billion in debt, becoming the company that carried the most debt globally, even though it wasn't even the biggest aerospace manufacturer at the time. That title belonged to the producer of the famous Super Hornets, McDonnell Douglas.

The Apollo Project was a product of the Cold War, while the Boeing 747 was the result of a bet between Bill Allen, president and a member of the board of directors of Boeing, and Juan Trippe, chairman of the board of directors for Pan Am, a very successful airline at that time. In October 1958, a Pan Am 707 airliner left New York for its first scheduled flight to Paris, which marked the beginning of a new era for jet plane travel. Jet planes not only flew fast but could maintain cruise altitudes above 32,000 feet, carrying them "over" the weather and making the flights easier. Trippe predicted that more than 35 million people would board international routes in 1965, and this number would grow by over 200 percent in 1980. As legend has it, in August 1965, Juan Trippe and Bill Allen were fishing on the Puget Sound near Seattle. Trippe asked Allen to build a jet plane that was 2.5 times bigger than the 707 while lowering the per-seat cost by 30 percent. Trippe said, "Would you build it if I bought it?" "Would you buy it if I build it?" replied Allen without hesitation.

As a matter of fact, the consensus among the aerospace manufacturers was to build supersonic airplanes. In 1962, the French and British governments signed a treaty to jointly finance and build the Concorde, a supersonic airliner. In 1963, the Soviet Union started the design of the Tu-144, and in the same year, Boeing set out to design supersonic jet 2707. In Joe Sutter's words, most people thought that the 747 was just a temporary solution, and once the supersonic airliners were commercialized, 747 planes would be replaced immediately.

Sutter was asked to leave the projects for the 707 and the 737 to become the chief engineer for a new airplane, which later became the Boeing 747. At the time, Sutter might not have been the best engineer in Boeing. He grew up in Seattle, and he could see the test airfield from his doorstep. From a very young age, his dream was to design an airplane to take off from that field. Even though Sutter had a strong team for the 747, nobody thought it would play any significant role in Boeing's future. Most people believed the hottest technology of that time, supersonic technology, would dominate in the future. Consequently, Boeing put its best technicians into the supersonic 2707 project, not on the 747.

Since Boeing wasn't sure about the potential of the 747, they thought it shouldn't be designed just for passenger routes but also as a freighter to lower the overall investment risk. During this period, airport congestion was a major problem. It would have been virtually impossible to load and unload all the cargo from a full airliner the proposed size of the 747 within the time frame that airports set. That's where the idea of being able to unload from the front of the airliner as well as from behind came from, thus the design of the raised cockpit to create a hump for the front-loading hatch. That design compromise turned into a signature of the 747, one that people are used to seeing today.

On December 22, 1965, Pan Am and Boeing announced an order of twenty-five of the 747 aircrafts for $525 million. In March 1966, the Boeing board of directors decided to proceed with the 747 program. On February 9, 1969, Boeing flew its 747 model for the first time. To launch the Boeing 747 in just under three years took an engineering miracle, since many of the technologies used had previously existed only in the dreams of the engineers.

The maximum thrust power for an engine was around 15,000 pounds at that time, while the 747 airliners needed to generate lift of more than 40,000 pounds. In order to take off normally, they had to wait on Pratt & Whitney's JT9D engines, which were then being developed, in what was one of the biggest gambles in the history of the aerospace manufacturing.

Before the Boeing 747, nobody thought an airliner could have two aisles, though today people take it for granted. That wasn't the case fifty years ago. It was the Boeing 747 that made two aisles on a single airliner a commonplace design feature.

Compared to the 707, the company's previous model, the Boeing 747 was twice as big. Even the enormous Airbus A380, which came out years later, was only one-third bigger than Boeing 747. The wings' three-part flaps increased surface area by 21 percent and lift by 90 percent. Around 75,000 engineering drawings were produced by a team of around one hundred people without the help of any computer-assisted design (CAD) software. Even today, with powerful tools like CAD, it's normal to deliver an airliner three years behind schedule. By contrast, the delivery of Pan Am's Boeing 747 order was on time without advanced digital tools. The Boeing 747 had around 4.59 million parts and 136 miles of electrical wiring. The Boeing 747 had lowered the per-seat cost of the airliner by 30 percent to a point where the cost of a Boeing 747 that could carry up to four hundred passengers was about the same as that of Concorde, which could carry only a hundred passengers.

On January 22, 1970, Pan Am's Boeing 747 made its maiden flight from New York to London. Compared to the cruise ships that crossed the Atlantic in from five to seven days, it took Boeing 747 airliners a mere eight hours, which made intercontinental travel possible for average citizens.

Unfortunately, Pan Am bought too many of the Boeing 747 airliners during the 1970s, so that when the Middle East oil crisis hit and fuel prices rose tenfold, it was devastating for both Pan Am and Boeing. Pan Am wasn't able to recover from the crisis and went bankrupt in 1991. Fortunately, Boeing gradually recovered in the 1980s. In 1989, the Boeing 747-400 was delivered for the first time, and that marked the beginning of another legendary chapter of its long history. Nearly seven hundred were delivered in just ten years, making the Boeing 747-400 the de facto Queen of the Skies.

In the meantime, the Concorde airliner, which test flew in 1969, wasn't put into use until 1976, when British Airways and Air France launched trans-Atlantic passenger routes between London and New York and Paris and New York, respectively. Only sixteen Concorde airliners were ever built, and in 2003 they had to shut down. Now Airbus might even stop building A380s, as they were able to deliver only 319 airliners, far short of the target of 1,200, which means Airbus is far from breaking even on this project. Nobody makes a big deal about flying in a Boeing 747 nowadays, and few remember what kind of sacrifice Pan Am made. There are around five hundred Boeing 747 airliners still in service today. They won't disappear from the sky in the near future.

I have been wondering whether we really hold technology in awe and why companies suffer different fates, though they worked on the same advanced technology. When we speak of our passion for technology, sometimes we don't realize what the cost might be. This might not be the best metaphor, but the relationship between technology and us is like the story of the farmer and the viper in Aesop's fables. It is told that, "One winter a Farmer found a Viper frozen and numb with cold, and out of pity picked it up and placed it in his bosom. The Viper was no sooner revived by the warmth than it turned upon its benefactor and inflicted a fatal bite upon him; and as the poor man lay dying, he cried, 'I have only got what I deserved, for taking compassion on so villainous a creature.'"

What does the passion for technology really mean? Do you truly believe that the technology will change the world? Do you have enough confidence and passion to warm up the viper in your bosom as the farmer did, even though it means you might be bitten after the viper wakes up? When you invest a large amount of passion into something, it's difficult to predict the final outcome.

When you place the viper within your coat, the biggest risk you might face is to your heart. You don't know if the viper will bite you when it wakes up, or whether your body is strong enough to handle it. Pan Am wasn't able to withstand it, and Mallory faltered on the route to the mountaintop. Everyone wishes the viper will be Madame White Snake in the "Legend of the White Snake," who represents a beautiful, kind woman in the Chinese folktale and came down to the Earth as directed by Buddha. But on the way up Mount Everest, Mallory was swallowed; on the journey to the moon, the

Baikonur launch site was the end for the Soviet cosmonauts, who remained earthbound forever. The open sky didn't offer the Concorde enough room to fly, and they too halted halfway to their dreams. The space shuttle program of the United States came to end officially on August 31, 2011, and wasn't able to meet up with the beautiful Madame White Snake as dreamed at the first launch, on April 12, 1981. Even though the road ahead looks bumpy and full of pitfalls, we still need to keep our hopes high.

We all know the saying, "Genius is 1 percent inspiration and 99 percent perspiration." However, in fact, Edison expended 200 percent in perspiration for his inventions and worked like a madman.

Interestingly enough, in 1907, a store selling shirts published an advertisement: "We carry the elements of success under our hats: 1 percent luck, 2 percent inspiration and 97 percent perspiration." Like all avid mountaineers, Mallory put in 200 percent of effort and sacrifice, yet he wasn't able to reach the top of the Mount Everest, for what he lacked was the 1 percent luck. We should hold that 1 percent luck in awe and gratitude, too.

City Brain: Moonshot Initiative for the Next Decade

Artificial intelligence and machine intelligence

In December 2016, at UCCA Center for Contemporary Art in Beijing, I had an interesting public conversation with Raffaello D'Andrea, a pioneer of drones and cofounder of Kiva Systems (now operating as Amazon Robotics). The conversation was about intelligent hardware. Dr. D'Andrea is a professor of dynamic systems and control at ETH Zurich. In his TED Talk in 2016, he showed an impressive demo of drones flying like fireflies. During the discussion, I mentioned the importance of differentiating artificial intelligence and machine intelligence, which was my first time sharing such an opinion in public. I think the definition of artificial intelligence should change over time. The idea of artificial intelligence used to be to make a machine think like a human. However, today artificial intelligence is asking the machine to do what human cannot do. It is more suitable to call today's artificial intelligence machine intelligence.

In May 2017, at the big data expo in Guiyang, I mentioned again that machines can do what human cannot and shouldn't do.

> Before people started using terms like "machine intelligence" and "artificial intelligence," there were only two types of intelligence: human intelligence and animal intelligence. Actually, anything that mimics human intelligence or animal intelligence can be called artificial intelligence, because humans created such intelligence. Together, that makes three types of intelligence: human intelligence, machine intelligence, and animal intelligence.
>
> The intelligence that aims to mimic animal and human can be called artificial intelligence. But if a machine has some intelligence that is beyond animal or human, this is the machine intelligence. That is my personal opinion.
>
> Today, when people speak of artificial intelligence, they're often concerned about whether human intelligence will be replaced by it. This is a rather unusual way of thinking. For example, when we use dogs to sniff for drugs, no one ever says that human noses are being replaced by dog noses. This example demonstrates that

people don't show enough respect to the abilities of machines. When machines exceed human capabilities in some respects, we should respect the achievement.

In August 2017, at the Global Artificial Intelligence Summit at Hangzhou, Yuval Noah Harari, the author of *Homo Deus*, who received a PhD in history from Oxford University, spoke about the relationship between artificial intelligence and artificial consciousness. His views made me feel strongly that we do need to differentiate artificial intelligence from machine intelligence.

Following Yuval's speech, I gave a speech on "Machine Intelligence, City Brain, the Apollo Program for the Next Decade." In my talk, I responded to Yuval's speculations that humans have capabilities beyond machines, such as the abilities of the unconscious mind. My point was that it's meaningless to compare humans with machines. I believe that, as technology advances today, we need to use machines to solve the problems human cannot solve, and to describe that the term *machine intelligence* is more accurate than *artificial intelligence*.

In the summer of 1956, a small group of scientists participated in the Dartmouth Summer Research Project on Artificial Intelligence, a workshop widely considered to be the founding event of artificial intelligence as a field. Four of the participants later won Turing Awards, and one of them was Herbert Simon. Simon visited Hangzhou a few times in the 1980s. He gave a talk about artificial intelligence at the Department of Psychology in Hangzhou University. At that time, I was still a student who had never stepped out of China before. I got very excited after listening to his talk, and I was convinced that artificial intelligence was the future! I didn't know until a few years later that Simon also didn't like the term "artificial intelligence" at first. Simon was a psychologist who had been awarded the Nobel Prize in Economics in 1978. Because of his visit, I made an important decision when I was about to graduate. Psychology can study humans from the perspective of biology or the perspective of information science. I chose information science, which required that I learn about computer science. From that point on, I started down a different path.

When people speak of artificial intelligence, they always talk about how to make machines do what human can do. In the famous Turing test, the fundamental idea is how to make machines mimic humans. But there is a paradox: does the human brain have the ability to understand itself? This is a still controversial question.

In 1981, Japanese engineers created fifth-generation computers with an information processing system, hoping Japan would take the lead in the global information industry. From the early 1980s to the late 1990s, artificial intelligence became popular. When Japan first announced the report from the fifth-generation computer committee, they invited Edward Albert Feigenbaum, professor at Stanford University, the creator of expert systems, and a student of Herbert Simon, as the committee member. Feigenbaum later wrote the book *The 5th Generation: AI and Japan's Computer Challenge to the World*. Some important observations in the book might originate from his wife, who is Japanese.

In 1980s, Feigenbaum coauthored *The Handbook of Artificial Intelligence*, which became my textbook to learn artificial intelligence. Its key thrust is about knowledge engineering, which is like the deep learning of today. This was also the first time I read such a heavy book, consisting of four volumes.

In the late 1980s, China proposed a national plan for an intelligent computer.

Unfortunately, in 1992, the fifth-generation computer project was officially shut down, and Japan permanently stopped at the so-called fifth-generation stage. The artificial intelligence I believed in wasn't going to happen. It was at this time that I started conducting research instead in virtual reality, which was part of the intelligent computer plan. Later on, the whole intelligent computer project transitioned from an "intelligent computer" to "computer software and hardware technology."

Today, artificial intelligence has come into the spotlight again. This isn't solely because of big advances in the field of artificial intelligence but because of changes of the world. One token of this was an article entitled "The End of the Internet, the Rise of Artificial Intelligence" published in the *Financial Times* in China. While I'm not sure whether we could really say artificial intelligence is on the rise today, I'm certain it is totally wrong to claim that the internet is ending.

Today, it is possible to talk about artificial intelligence again because of three key elements that are the primary topics of this whole book.

The first is the internet. I've argued that the internet has become the most important infrastructure in human history. If we didn't have this infrastructure, we wouldn't have data. As an infrastructure, the internet can't be confused with the traditional network. Personally, my biggest dream is to see the human brain connected to the internet.

The second key element is data. The advances in speech recognition today are based not only on deep learning models but more importantly on the data. Without an infrastructure like the internet, we wouldn't have the breakthroughs made in speech recognition. The same goes for the breakthroughs in imaging technology. What does imaging technology provide us with? Because of advances in imaging technology, people have a new hobby: the selfie. Selfies as a hobby result in lots of data about the people themselves. In addition, self-driving cars are already using a lot of imaging technologies. No matter whether the cars is equipped with cameras or radar sensors, the key technology is imaging technology. Self-driving cars have become imaging-technology cars.

In July 2017, I visited several companies. One specializes in working with satellite image data. This company uses satellite images to estimate the reserves of oil in oil tanks worldwide, which is highly relevant for the oil futures market. Another company is a Canadian start-up. The founder told me that what his company specializes in is AI, but not artificial intelligence, affective intelligence. You use cameras not to do facial recognition but to measure heartbeat and blood pressure. You could conduct a physical exam of yourself every day while playing with your cell phone. Here already is something humans cannot do. In the same way, actually, the artificial intelligence claimed by many companies is already doing something beyond what humans can.

The third key element is computing power. There has been a breakthrough in computing power as well, and deep learning uses a lot of computer power. The basic unit of computing power is the number of transistors. I once heard a surprising figure, though it may not be very accurate: that a person today consumes 2 billion transistors every year. What does that mean? In the 1970s, an Intel CPU had only two thousand transistors. If we do some simple math, the total number of transistors in the United States in the 1970s may have been less than what a single person has today. Such an advance in computing power is beyond imagination. And this breakthrough became the basis for us to reexamine artificial intelligence. I visited a start-up in Silicon Valley focusing on inventing chip technology that aims to provide a thousand times more computing power than today's best GPU. From two thousand transistors to the two billion transistors in the Intel microchip, and then to a thousand times more computing power—what we can't do in a lifetime today may be achieved in the blink of an eye in the next decade.

We need to remember, silicon is fantastic, no matter whether there is a Moore's Law or not. In 1970s, there were only a couple thousand transistors in a chip. We have made enormous progress to get to billions of transistors per day per person. That is great progress for the world.

Artificial intelligence today mainly relies on these three key elements. They don't have to do with the understanding of human brain, which is why I believe it's better to call it machine intelligence. We should let machines do what humans cannot do. But I believe that one thing the vigorous discussion of artificial intelligence does indicate is, once again, the influence of the internet.

The Apollo program for the next decade

Today, the biggest challenge cities face worldwide is sustainability. This challenge brings with it great opportunity: how to use machine intelligence to solve problems such as traffic congestion within the city. This is a great opportunity for machine intelligence to make a breakthrough, which is one reason why I'm dedicated to City Brain with my whole heart.

In the past twenty years, China has been investing in building city information systems. Every city has accumulated a rich resource of data. However, without machine intelligence, it would take a person more than a hundred years to look through all the data about a day's traffic in a given city. Before they were fully utilized, such valuable data used to be quietly deleted forever, simply because there was no available storage.

In October 2016, the Hangzhou city government announced the City Brain initiative at the Apsara conference in Hangzhou, with the participation of dozens of companies. I gave a speech at the conference in which I observed that the longest distance in the world is not from the South Pole to the North Pole but the distance between the traffic signals and the traffic cameras. The traffic signals and cameras are on the same pole but they were never connected and never talked to each other—the data wasn't flowing between them. When the data isn't flowing, the traffic can't flow smoothly. Traffic signals act independently of the actual traffic conditions.

Building on the internet infrastructure, the purpose of City Brain is to use the city's rich data resources to better allocate public resources and enable sustainability. Problems such as traffic congestion are actually created by the poor planning by human beings. The core idea of City Brain is to acknowledge the importance of data as a resource for the city's development, which for the future is an even more important resource than land.

In Hangzhou, for the first time, the data from traffic cameras are fed back to the traffic signals, thanks to City Brain. The traffic signal control plans have improved traffic in the downtown area. Addressing traffic problems is just a beginning, though. The important thing is that data start to show its value in society. I believe City Brain will bring about three important breakthroughs for urban development:

First is a breakthrough in city management. We can solve the biggest challenges in this area and realize a novel human-centered management of cities.

Second is a breakthrough in city services. City Brain will be a foundation for the government to serve its people better, whether individuals or companies. Public services, such as transportation, will enter the age of effective and efficient service, avoiding the waste of public resources.

Third is a breakthrough in city industry. Open city data will become a fundamental resource for the transformation of traditional industries and for the development of new industries. It's just like how oil and semiconductors empower the oil industry and IT industries, respectively.

City Brain isn't just a technology innovation; it's similar to what New York did with electricity more than a hundred years ago. It speeds up a change in mindset from private data to open data. By building the neural networks of the city, City Brain is using the data to think in real-time, make decisions, and take actions.

There is no other city in the world that has a City Brain–like infrastructure. Thanks to the internet infrastructure in China, City Brain was first proposed and constructed in Hangzhou. Today, while most of families in Americans are still using checks to pay the water and electricity bills, people in China are already using cell phones to pay for their street food. This small difference shows the unique power of internet infrastructure in China.

In May 2017, I had a conversation with Zhang Han, a journalist from the Chinese monthly magazine *People*. Below is what she wrote in the final article, entitled "Jian Wang: The Destiny of a Prophet":

Cloud computing is very popular today.

Jian Wang no longer needs to argue with those who have doubts about cloud computing and do battle with them like Don Quixote.

Today, Jian Wang no longer has a role managing the company. He needs to broaden his horizon and try new endeavors.

Strangely, when communicating with government officials, he's no longer someone who is hard to understand. Now he demonstrates his high EQ. His strategy for city management is readily adopted by government officers.

His hope is that government departments will accept his vision for the future city and deploy a City Brain in the city. . . .

It may seem the future is far away. Jian Wang, however, believes, it can be realized in China. City Brain is as important as the power grid. China now has the opportunity to give the world a new infrastructure. . . .

It seems that he not only sees the future. Artificial intelligence is so hot that everyone wants a piece of it. Jian Wang remains as determined as ever, he sees it, and he will also make it happen.

Someone read his Chinese version of *Being Online* and left the comment: "You make the prediction and you make it happen. It's like cheating." Jian Wang feels the comment makes a lot of sense, "Let me continue to cheat then."

City Brain is not only important for the city sustainability development. It is also important for technological innovation. It provides a large enough field for people to explore through experimentation.

The Apollo Program in the 1960s generated new technologies, such as telecommunications and bioengineering. Today, when the internet has already become the basic infrastructure in China and the rest of the world, we have an unprecedented amount of computing power and data resources, and City Brain will not only benefit the people living in the city but will generate new technologies. Just like Apollo, City Brain will become the most important innovation platform in the next decade.

Every technology revolution is an important step for civilization. In the age of the steam locomotive, the development of a city was measured by how many roads a city builds; in the age of electricity, the development of a city was measured by how wide the coverage of the power grid. In the age of the internet, data has become an important resource, and the city will build a new data infrastructure to take it to another level. Just as London operated the first subway trains 160 years ago and New York first built power grids 135 years ago, I hope City Brain will become a brand-new city infrastructure—a technology gift from Hangzhou to the world. In my speech at the Annual National Planning Conference in 2017, I envisaged that in the future a city with City Brain would consume less than 10 percent of the natural resource (in water, electricity, and land usage) as it needs today. An online city will make this possible.

This is just the beginning for the online era that is the combination of the internet, data, and computing, and the 1 percent opportunity is absolutely tempting. The mountaintop for data in the online era is just as enthralling as Mount Everest yet also as challenging and unforgiving. Computing in the era of being online could send people to the moon with rocket science, but without the serendipitous 1 percent it could also result in smoldering husks of rockets like what we saw at the Baikonur launch site. The internet in the new age could allow you to move freely among the internet of things like the Boeing 747 in the skies, but it could also trap your dreams within a destination triangle of London, Paris, and New York, like the beautiful Concorde.

Regardless, we are on a path to create technological innovation that is unprecedented. See you at the finish line!

Epilogue

In the New Year of 2018, I received a book from one of my friends, titled *You Say to Brick: The Life of Louis Kahn* by Wendy Lesser. It is the first biography of the iconic American architect Louis Kahn (1901–1974). One of the greatest architects of his era, he once said: "The world never needed Beethoven's Fifth Symphony until he created it. Now we could not live without it."

Indeed, there are quite few inventions without which we could not live.

In November 2013, on the plane to Taipei, I browsed the November 25 issue of *Time* magazine. There was an article by Jeffrey Kluger, "The Spark of Invention." The contents of the article weren't that exciting, but two points in it are very interesting. One is that the three most important inventions in history are electricity, the internet, and the wheel: in themselves, electricity and the wheel are useless, but if there were no electricity, there wouldn't be televisions, refrigerators, computers, etc. If the wheel weren't part of bicycles, cars, trains, even airplanes, it would have no value in itself. The same is true for the internet. Second, the survey found that 84 percent of people around the world think this is the best era for invention. A story goes that in 1899 the head of the US Patent Office, Charles H. Duell, announced that "everything that can be invented has been invented" and wrote to the then US president to close the patent office. This legend has never been verified, but did it indirectly reflect the public's opinions at the time? In contrast, is today's public clairvoyant or blind? If we agree that electricity enables all the inventions of electrical things, that wheels enable all the inventions with wheels, we should have more faith in believing that, as we have only recently entered the era of being online, we will see even more inventions during this period.

There is a description concerning the design of wheels in the ancient Chinese book *Zhou Li · Dongguan Kao Gong Ji*. From the requirements in the book, we can see the significance of the wheel design of cars today and how the way they're designed is important for the car's adaptability to different road conditions. It so happens that in 1975 an Eastern Han Dynasty copper light carriage (a carriage with double shafts and a single horse) was unearthed in Wantun Town, Xingyi City, Guizhou Province. The carriage also had large wheels with small hubs, which was more suitable for the five-foot road that ran between Yunnan and Guizhou at the time. I think the phrase "Do not reinvent the wheel" could be interpreted in another way: the wheel in itself creates additional opportunities for invention in this world.

Similarly, we don't need to reinvent the internet today. The internet of things that is so much talked about now and the internet of vehicles are all part of the internet. This is why I gradually understood that the mobile internet is still the internet. I once joked that the internet offered by Google Search is only online news and published websites. The online world is much bigger than what Google presents to us. In the online era, we face more uncertainties. If there is one thing that is certain, though, it is that we will certainly create more things that people cannot live without.

We're lucky to be living in a time where creation may determine the future of this era. But whether creation will determine the future depends on our belief and persistence. Believe in persisting, and persist in believing.

JIAN WANG
YUNQI, JULY 2021

Acknowledgments

The support of many people made this book possible. In particular, my thoughts and opinions have benefited from conversations and discussions with many bright people, and I want to express my gratitude to them. Special thanks to Yuefei Pan, Jingyi Li, Jiang Liu, Xiangming Liu, Yuting Zhang, Jiashi Li, Xuemei Guo, Guohui Zhou, Jun Xu, Qian Li, Yisi Wu, Lei Wu, Zhigang Chen, and many other friends who have spoken with me. Their creativity and vision of the internet gave me different perspectives on some of my basic ideas. I also thank them for letting me use the interview materials that appear in this book. Additional thanks to Zhenhui Li for her great help during my writing of this English version. Last but not least, many thanks to Calvert Barksdale, this book's editor in English; his questions and editing helped me improve the book greatly. Special thanks to Weihua for her patience and tolerance during my writing of this book.

Permissions Acknowledgments

Index